RALEIGH IN EXETER
1985

Privateering and Colonisation in the reign of Elizabeth I

Papers delivered at a conference at Exeter University on 3–4 May 1985 to mark the four hundredth anniversary of the first attempt to settle English people in North America.

edited by

Joyce Youings

EXETER STUDIES IN HISTORY NO. 10

UNIVERSITY OF EXETER

First published 1985 by the University of Exeter
© Department of History and Archaeology, University of Exeter 1985
ISBN 0 85989 252 2
ISSN 0260-8626

EXETER STUDIES IN HISTORY

General Editor: C. D. H. Jones, BA DPhil FRHistS

Editorial Committee

Professor I. A. Roots, MA FSA FRHistS   M. D. D. Newitt, BA PhD
M. Duffy, MA DPhil FRHistS   B. J. Coles, BA MPhil FSA

Publications

Printed in Great Britain by A. Wheaton & Co. Ltd, Exeter

CONTENTS

LIST OF FIGURES AND PLATES

Figures

Plates

Introduction

Joyce Youings

On 9 April 1585 there sailed from Plymouth an expedition master-minded, and probably largely financed, by a Devonian, Sir Walter Raleigh. It comprised seven ships and was under the general command of Sir Richard Grenville, of Stowe in the Cornish parish of Kilkampton. Besides their crews these ships carried fourteen gentlemen and nearly one hundred other men, all of whom were landed some weeks later on the south eastern shore of North America. There, on Roanoke Island and the territory thereabouts, they were to remain for nearly twelve months, the first English people to over-winter on the North American mainland. They were not venturing onto entirely unknown territory for a year earlier Raleigh had sent a reconnaissance expedition to the area. Philip Amadas and Arthur Barlowe, both young Devonians, had returned with glowing descriptions of the attractions for English settlement of what is now the coast of the state of North Carolina. Their work was followed up in 1585–6 by that of the Cornish artist, John White, and his mathematician colleague, Thomas Harriot, who drew and mapped, in collaboration, the land Raleigh called Virginia. But settlement did not proceed smoothly: the first colonists returned home with Sir Francis Drake in 1586. In 1587 a second expedition sailed, this time carrying a handful of women and children, with White as their governor. Deserted by a homeland preoccupied with preparations to defeat the Spanish Armada, all except White (who had returned home to fetch more supplies) disappeared, virtually without trace, the so-called 'Lost Colony'. All would seem to have been in vain but enough had been learned of prospects for North American settlement to form the basis for later and more enduring enterprises.

The people of North Carolina, and especially the citizens of Raleigh, the present state capital, are very proud of their English connection and are celebrating the *Quadricentennial* with tremendous enthusiasm, enterprise and no little expenditure of public and private funds. In England the most important contribution to the celebrations has been the British Library's major exhibition, 'Raleigh and Roanoke'. Staged first in London it has been taken across the Atlantic, first to the city of Raleigh and then to New York.

On 3–5 May 1985 a conference was held at Exeter University under the combined sponsorship of the university's departments of History and Archaeology and of Economic History, the Devonshire Association, the Exeter branch of the Historical Association and the Exeter and East Devon branch of the English Speaking Union. A grant from BAT Industries made it possible to invite Dr H. G. Jones, Curator of the North Carolina Collection in the library of the University of North Carolina at Chapel Hill, a distinguished 'Tar Heel' and a former Keeper of the State Archives. He spoke on 'The Americanization of Raleigh'. Included in the conference programme was the University of Exeter's Harte Lecture for 1985, given by Professor Kenneth Andrews of Hull University on 'Elizabethan Privateering', a subject, as he showed, very closely related to his recent book, *Trade, Plunder and Settlement* (Cambridge University Press, 1984). Mr Ian Friel of the National Maritime Museum at Greenwich set the Tudor maritime scene with his paper on 'The Three-masted Ship and Transatlantic Voyages', showing the importance of technology, including that of English shipwrights, in New World exploration. The conference was then asked to consider the question, 'Did Raleigh's England need Colonies', an attempt to compare the claims of promotional literature with the realities of contemporary society. Professor David Quinn of Liverpool, the leading authority on the early settlement of North America, whose most recent book, *Set Fair for Roanoke* (University of North Carolina Press, 1985) is a wonderfully fresh and imaginative re-telling of the story, spoke on 'The Lost Colonists', concentrating on their success and failure to achieve a *modus vivendi* with the native inhabitants. Dr Michael Stanford of Bristol and the Rev. Maurice Turner of Leaton near Shrewsbury spoke, respectively, on the two Walter Raleighs, father and son, and succeeded in showing that not only was the latter a remarkable man but that he was a member of a remarkable family. Finally Mr Paul Hulton, lately deputy keeper of the Department of Prints and Drawings at the British Museum, talked, with excellent slides, on the subject of his recent book, *America 1585, The Complete Drawings of John White* (University of North Carolina Press and British Museum, 1984). The sponsors of the conference were honoured by the presence in Exeter of so many of the leading scholars in the field of Elizabethan overseas enterprise who included, besides those already mentioned, Dr Helen Wallis of the British Museum Map Library, principal author of *Raleigh and Roanoke*, the catalogue of the exhibition already mentioned, and Professor Karen Kupperman of the University of Connecticut, author of *Roanoke: the Abandoned Colony* (Rowan and Allanheld, Totowa, New Jersey, 1984). The conference ended very appropriately with a preview and discussion, led by Dr Stephen Fisher of the department of Economic History, of plans for the four-year investigation of the maritime history of the county of Devon, a project based on the University of Exeter which is being funded by the Leverhulme Trust. A version of Dr Fisher's paper has already been published in the *Mariner's Mirror*.

All the speakers have generously agreed to the publication of their papers. With the addition of footnotes these have been printed very much as they were delivered. In several cases, and without prior arrangement, themes touched on in one paper were taken up by later contributors. Careful readers will detect minor differences of opinion but, perhaps disappointingly, there was no great 'debate', even about privateering. So great is the current interest in the achievements of Raleigh's servants, especially White and Harriot, that there is a danger that Raleigh himself, who, after all, never actually went to Virginia, is in danger of being forgotten. But implicit in all the papers is a recognition of his indispensable role, particularly in 1585, as the promoter of the first attempts at English overseas colonisation. Perhaps, before we become too preoccupied with the Armada anniversary, the time is ripe for a reappraisal of the man who, even more than Sir Francis Drake, is the foremost Anglo-American hero. There is, incidentally, no particularly Anglo-American dispute over the spelling of his name. He himself spelled it in many different ways, but current practice among historians seems to be moving in favour of 'Ralegh'. Contributors to this volume have been allowed to follow their own fancy, if only because of the editor's devotion to the traditional spelling of the name of the small hamlet in north Devon, less than a mile from which she herself was born, where the Raleigh family had its beginnings.

The last word on the Roanoke Voyages is far from being said. Each of the conference papers, while containing the results of the latest research, most of it by the authors themselves, offers pointers to the need for more research and to the possibility of further discoveries. The best hope of picking up the trail of the lost colonists would seem to be in the field of archaeology, and as yet historians know very little about where in England they came from. Though they sailed from Plymouth few are now thought to have been westcountry people. But the South West has a tradition, more perhaps in the twentieth century than in the sixteenth, of playing temporary host to visitors from other parts of the country and from overseas, and this conference enabled Exeter to honour Sir Walter Raleigh, both of whose parents are buried within its walls, by doing just that. It was a fruitful and enjoyable Anglo-American occasion.

August 1985

1

Elizabethan Privateering

The Harte Lecture 1985

Kenneth R. Andrews

I wonder whether people these days read Charles Kingsley's *Westward Ho!* I don't suppose many do, even as children. I can't remember whether I did so myself as a boy—it's rather a long time ago—but since I became interested in privateering I have, for my sins, taken the trouble to peruse its purple pages. It was published originally in 1855, and went through many editions between those dates. My guess would be that it was even more popular in the 1920s than it had been in Victoria's reign: but admittedly that is only a vague impression based on general reading. What is needed, and could be a fascinating study, is a thorough investigation of how this book and ideas of that sort were received and imbibed (or not, as the case may be) over the whole period since 1855, and indeed whether and how far it is possible to trace the ideas back—even perhaps to Elizabethan times, though naturally one would expect ideas to change their form and their thrust through so many generations.

For it is the relationship between the romantic image and the sordid reality which is to my mind at present the most interesting and significant aspect of the subject. To get the flavour of Kingsley, let me quote you his dedication of Westward Ho! addressed to two Victorian pioneers of empire.

> That type of English virtue, at once manful and godly, practical and enthusiastic, prudent and self-sacrificing, which [I have] tried to depict in these pages, they [that is the two Victorian dedicatees] have exhibited in a form even purer and more heroic than that in which [I have] drest it, and than that in which it was exhibited by the worthies whom Elizabeth, without distinction of rank or age, gathered round her in the ever glorious wars of her great reign.

As the introduction to the 1924 edition tells us, 'Westward Ho! makes vivid the spirit of eager adventure and the pride of patriotism which made the tiny country of England so astonishingly successful against her huge and angry neighbour, Spain.' And the hero, Amyas Leigh, was a very real English youth of those stirring times, such as was pictured in that well-known Victorian work of art: 'The Boyhood of Sir Walter Raleigh'.

Is it fair to cite Westward Ho! as typical of English writing about Elizabethan privateering? Perhaps not: after all, it was a novel, not intended for adults and hardly comparable to history proper. But Kingsley was a historian. He was professor of modern history at Cambridge from 1860 to 1869 and was especially well informed about Elizabethan maritime history, the West Country and the Caribbean.[1] He seems to have absorbed a great deal of Hakluyt's Voyages, and though he deliberately mixed up fact and fiction he also grounded his plot and his characters carefully in the Elizabethan world as he understood it. The trouble is, of course, that he didn't understand it, and that his interpretation has contributed to many other people's illusions, partly because it was so well done, partly because it appealed to children and partly because it said what most English people were quite happy to believe, and left out what they didn't want to know. And I submit that the popular literature of the entire field of Elizabethan maritime expansion and even much of the straight history on this subject have shared—and still largely do share—the nationalistic, sentimental and historically false conceptions propagated by Kingsley. I don't mean that they necessarily derive them from him, but that he and they look through the same sort of spectacles. Even that Cornishman A. L. Rowse takes the same romantic, if rather more sophisticated, view. As far as this subject is concerned he is in the mainstream of the nationalist tradition. Look for instance at the way the traditionalists characterize privateering and its relation to piracy. The war with Spain, as every schoolboy never knows, began in 1585, in the sense that ventures of reprisal against Spanish shipping were legally authorised from that date. Before then such actions, like those of Drake, Oxenham and company, were piratical. Nevertheless they are commonly described as privateering in this literature, and even Rowse, who knows better, does the same. Kingsley hardly mentions piracy, though he was in fact writing mainly about pirates, endowing them with all those virtues of godliness, manliness and patriotism in the struggle for the true religion and the freedom of the seas.

But what most of these writers ignore is that before 1585 Englishmen were responsible for criminal and indiscriminate piracy on a disturbing scale, the menace having increased to such proportions since the middle of the century as to become a major headache for Elizabeth's Privy Council. Many hundreds of men in those years were convicted of piracy and there is no doubt that thousands more actual pirates were never convicted, for the problem was simply unmanageable. Godliness, manliness and patriotism had very little to do with it. It was simply the maritime version of robbery with violence, which in that context often meant killing some of the victims and torturing others. Although quite small scale at this stage—usually a matter of one or two ships—piracy was very widespread, basing itself in the minor ports and creeks of the East and South coasts, the West Country, Wales and Ireland: robbing barks and small craft chiefly in the North Sea, the Irish Sea and the Channel. And of course it had substantial backing ashore—it could not have continued otherwise. Gentlemen great and small were frequently interested parties, admiralty officials especially, and discreet merchants were involved in the disposal of the stolen goods. We have several detailed studies of piracy, mostly concerned with the vain efforts of government to suppress it. It is clear that piracy was on the increase—and the government increasingly worried—from Elizabeth's accession right down to 1585, when it was largely subsumed in the privateering war. In fact the notion that English seamen took to plunder only under extreme provocation and from the highest motives is myth. Some of course did have specific grievances against Spain: most notably the Hawkinses, who set forth Drake in private reprisal for losses suffered at Spanish hands in their last slaving venture to the Caribbean. But even this famous case should be seen in the context of that endemic piracy in home and nearby waters in which the Hawkinses of Plymouth had taken a leading part for years before their traumatic experience at Vera Cruz. And the majority of those who attacked Spanish shipping in the decades before 1585 had no such excuse. Let me hasten to say that I am not suggesting that English piracy of this ordinary, endemic and criminal type caused the war with Spain, or even that it was an important factor in bringing about that war. Indeed I don't believe that English efforts at overseas expansion in general—commercial, predatory, colonial or whatever—played any great part in causing it. I am only saying that the pre-war piracy was not privateering, nor was there normally anything noble, patriotic or religious about it: rather the reverse. On the other hand it may be worth looking at it a bit closer if we wish to understand privateering, since privateering did in part grow out of it and inherited some of its features.

Why was it so widespread and why was it growing? Contemporaries well placed to judge the matter were generally agreed about the root of the evil: the poverty of the common seaman. As Hakluyt wrote: 'the present short trades causeth the mariner to be cast off, and oft to be idle, and so by poverty to fall to piracy.' Sir Henry Mainwaring, once a pirate himself, also ascribed it to the fact that 'the common sort of seamen are so generally necessitous and discontented.' Sir John Hawkins and Captain John Smith said much the same thing.[2] Elizabeth's reign was a period of commercial instability: trade was repeatedly interrupted, the whole pattern of international trade was disrupted by the decline of Antwerp, new trades were beginning, but old and important ones were in difficulties. Unemployment among seamen was chronic and wages poor, at a time when the population was increasing fast, especially in the ports, and wages generally were lagging behind prices. At the same time the maritime circumstances and the conditions of sea warfare provided sailors with frequent and tempting opportunities for spoil and pillage. When private disputes arose between merchants or seamen in different countries they often led to local reprisals and counter-reprisals between the ports concerned, festering and proliferating for years. A long back-log of unsettled Anglo-French disputes for example was the subject of international negotiation in the mid 1580s. Major international disputes would produce trade embargoes accompanied by a spate of reciprocal depredations frequently involving neutrals and tending to persist after the governments concerned had come to terms. In the wars with France which punctuated the middle decades of the century many of those who undertook private operations against enemy shipping were unlicensed, particularly when, as in 1544 and 1557, permission was granted to all and sundry by proclamation. The wars with Scotland at this time gave similar opportunities for unlicensed privateering which, because it was uncontrolled, could easily deteriorate into piracy as promoters, captains and crews sought to make their voyages pay, while some, naturally enough, continued this less discriminating business after the cessation of official hostilities.

In such ways were ordinary seamen inducted into disorderly and predatory cruising, because of course the ships involved were usually the common or garden merchantmen of the time, more or less adapted for purposes of plunder by the addition of men and guns. This sort of endemic violence, whether in the shape of petty, even casual marauding or of professional piracy pursued by notorious and widely-feared malefactors with private and well-armed men-of-war, developed rapidly in the early years of Elizabeth's reign and was soon aggravated by the civil wars in France and the rising of the Dutch rebels against Spanish rule. Huguenot ships from La Rochelle and the so-called 'sea-beggars' from Holland and Zeeland frequented English ports, recruiting English sailors. English captains accepted dubious letters of marque from William of Orange or Gaspard de Coligny, the Huguenot leader, and for three or four years around 1570 a state of near-anarchy supervened in the Channel and nearby waters, endangering all shipping in the area, in the course of which time this storm of violence spread into the Atlantic, past the Canary Islands and across to the Caribbean.

It is easy enough to blame the government for all this. It not only encouraged shipowners to set forth against the enemy in time of war: it issued commissions indiscriminately and exercised no serious checks on how they were used or abused. Even in times of nominal peace it allowed Englishmen to serve foreign princes with their ships and turned a blind eye to acts of illicit spoil whenever it was politically convenient to do so. Moreover the attitude of the state to piracy was ambivalent. It condemned it when trade suffered, allies were offended or influential English merchants protested, but it was quite prepared to adopt pirates to serve itself. The Yorkshireman, Martin Frobisher, was an outstanding instance of this in the sixties: he was arrested three or four times on piracy charges but ended up being employed by Cecil as a ship's captain on government business. A few years later when he was trying to get subscriptions for his northwest passage project he found it hard going because merchants suspected that once provided with ships, men and supplies he would return to his deplorable habits. Sir Francis Walsingham, as the queen's secretary, took under his protection the notorious Portuguese pirate, Simon Fernandes—he who was to be the pilot of England's North American pioneering in the 1580s, leading, among other voyages, all the main expeditions to Roanoke. But these are only two instances of many I could cite, including of course Francis Drake. As Mainwaring, having risen from pirate to admiral, put it: 'the state may hereafter want such men, who commonly are the most daring and serviceable in war.'[3] But in a sense the government had little or no choice. It had not the resources to build or provide a professional navy sufficient to fight its wars and therefore depended on private shipping to harrass enemy commerce or make a nuisance of itself at other times. And it lacked the power to put down piracy when it got out of hand. That would have required regular patrolling, which the crown could not afford, and effective action against the patrons of pirates, which was not impossible, but difficult to carry through systematically, since the protectors of pirates included vice-admirals of the coast like Sir John Killigrew, and men of much greater power like Sir George Carey, captain-general of the Isle of Wight, vice-admiral of Hampshire, brother-in-law of the Lord High Admiral and kinsman of Queen Elizabeth.

I certainly don't want to suggest that every sailor involved in such business was a dyed-in-the-wool villain, but I do think it is unrealistic and tendentious to present them as risking their lives for the queen, the country and the true religion. Of course motives were mixed and were interpreted in different ways at the time. Richard

Madox, for example, who went as chaplain aboard an expedition of 1582 intended for the East Indies, a trading venture which turned, as many did, to plunder, kept a secret record of the voyage. When the first piratical attack occurred he preached against such conduct, only to be told by the culprits: 'we could not do god better service than to spoil the Spaniard of life and goods.' To which his own comment was: 'but indeed under colour of religion all their shot is at the men's money.' In any case the prize was not a Spanish but a Flemish vessel.[4]

But then again, it might be urged, all this petty, local piracy is beside the point. Kingsley and Froude, the Victorian prophet of empire, and Corbett, the author of the classic *Drake and the Tudor Navy*, and all those Edwardian beaters of Drake's drum, and glowing biographers of Gilbert and Ralegh: they were talking about the pioneers of the Atlantic and explorers of North America, dreamers and planners of English expansion overseas, the very first heroes of the British Empire. I will not stress that such writers had a fairly large axe to grind, in the shape of the empire of their day. Let us rather look at the historical facts. Drake was of course a pirate long before the queen engaged his services and he continued as a pirate in her service until he became rich. Then he did nothing until the outbreak of war, when he became a privateering admiral. He was plainly a great patriot and something of a religious fanatic too, but every single one of his ventures was devoted to plunder. He couldn't even refrain from going after a prize in the middle of the Armada action of 1588. This naturally enraged his brother-pirate Frobisher, who afterwards, we are told, uttered certain speeches, to wit:

> Sir Francis Drake reporteth that no man hath done any good service but he...he hath done good service indeed, for he took Don Pedro [i.e. Don Pedro de Valdés in the galleon *Nuestra Señora del Rosario*]. For after he had seen her in the evening, that she had spent her masts, then, like a coward, he kept by her all night, because he would have the spoil. He thinketh to cozen us of our shares of fifteen thousand ducats; but we will have our shares, or I will make him spend the best blood in his belly; for [I have] had enough of those cozening cheats already.[5]

In this profitable action, I may add, Drake was assisted by a ship called the *Roebuck*, belonging to Sir Walter Ralegh and commanded by Ralegh's faithful servant Jacob Whiddon.

But what about Ralegh and his half-brother Gilbert, the boys in that famous picture? As that picture suggests, they certainly grew up aware of the sea and its opportunities, for Ralegh's father was repeatedly in trouble with the Admiralty Court for piracy out of Exmouth: and duly achieved the appropriate office of deputy vice-admiral of Devon. This gentleman married Humphrey Gilbert's mother and Gilbert in his turn had a good deal to do with pirates, though he himself was no seaman. The queen actually at one point forbade him to undertake his last voyage on the grounds that he was 'a man noted of not good happ by sea'[6]—though she relented of course and Gilbert sailed on that disastrous expedition, in which he lost his flagship with most of his men and finally was drowned on the way home. In this venture one of his ships committed a particularly brutal and unwarranted attack on a French fishing vessel—a crude piece of piracy—but Gilbert had been much worse compromised in his first attempt at an American voyage in 1578. The fleet he prepared on that occasion seems to have attracted pirates in plenty, presumably because it was intended as a raid for plunder upon Spanish possessions, though there was no war or real threat of war at this stage. The upshot was that half Gilbert's force made no attempt to cross the Atlantic, but went cruising for whatever they could take, which meant in practice French shipping as well as Spanish property. It would be unfair to blame Gilbert directly for all this, but he was responsible for the conduct of the force as a whole and it is not clear that he was perfectly innocent. As to his intentions, some sort of plunder cruise was certainly envisaged, but otherwise they are wrapped in mystery, which is not much elucidated by the verses of the poet Thomas Churchyard saluting this journey, who, if I may quote:

> Marvelled how this knight/could leave his lady here/His friends and pretty tender babes/that he did hold so dear/And take him to the seas/where daily dangers are/Then weighed I how immortal fame/was more than worldly care/And where great mind remains/the body's rest is small/For country's wealth, for private gain/or glory seek we all.[7]

This eulogy is remarkable in two respects: in the first place it fails to mention God, which is most unusual; and in the second place it does mention private gain, which is also rather unusual. My impression of Gilbert is that he had no strong religious inclination, but was extremely ambitious, greedy, domineering and given to violent fits of rage, not to mention the murderous cruelties he committed in Ireland. He certainly does not fit clergyman Kingsley's concept of imperial virtue and muscular Christianity.

Ralegh is of course more complicated, much more important and very much more interesting. Edmund Spenser in the *Faery Queen*, which I suppose to be the most unreadable poem in the English language, described him as the 'shepherd of the ocean'. I have never been able to understand why. Is it supposed to mean that he aspired to protect those sheep of the ocean, the defenceless merchantmen? If so, Spenser was rather wide of the mark, for Ralegh was much more like a wolf on the ocean in that respect. Of course he did not have much opportunity to appear in the role of pirate or patron of pirates because soon after his star rose above the horizon, the long impending war arrived and clothed piracy in a mantle of legality, if not respectability. But at least in the role of privateering promoter, organizer of ventures of single ships, squadrons or huge fleets, and as a privateering captain, admiral and conquistador of the sea Ralegh easily qualified as a wolf in sheep's clothing: if you will forgive me flogging that dubious metaphor. This is a fact which I could substantiate in considerable detail, but I will only refer you to the opinion of one who was his unstinting admirer and practically his apologist, V. T. Harlow. He writes that 'one of Ralegh's chief sources of wealth was privateering, which he conducted as a regular business throughout his career' and relates a whole series of cases in which Ralegh, his captains and his associates committed illicit spoil of neutrals, embezzlement of prize-goods, bribery, fraud and outright piracy.[8] Yet this was the man who could say without a blush in the dedication of his book, *The Discoverie of the Large, Rich and Bewtiful Empire of Guiana*, 'It became not the former fortune in which I once lived, to go journeys of pickery ... to run from cape to cape and from place to place, for the pillage of ordinary prizes.'[9]

This was, to say the least, inexact, and few people at the time would have swallowed it, for Ralegh's prominence in that undignified scramble for loot was too obvious. But the disclaimer undoubtedly arose from two considerations. On the one hand Ralegh did not wish to be associated in public with the disreputable business of privateering. It did not suit the noble image he sought to present to the world. Sir Robert Cecil, his partner in various privateering operations, felt the same. As he actually wrote to Ralegh in connection with one such venture:

> I pray you as much as may be conceal our adventure, at the least my name above any other. For though I thank God I have no other meaning than becometh an honest man in any of my actions, yet that which were another man's pater noster would be accounted in me a charm.[10]

The two of them knew they were soiling their hands, but didn't want everyone to see the dirt. Secondly, moreover, Ralegh was anxious to present the Guiana enterprise as a piece of empire building rather than the treasure hunt most people took it to be. His argument was that in Guiana lay the strategic key to the defeat of Spain; he had not undertaken it for private profit but for the queen; he had refrained from plundering the Indians in order to win their alliance in a future campaign of conquest with royal backing. In short, he was not only a war-leader of rare strategic vision, but a knight in shining armour sacrificing his own meagre fortune for the benefit of the realm:

> If it had not been in respect of her highness' future honour and riches, I could have laid hands on and ransomed many of the kings and caciques of the country...but I have chosen rather to bear the burthen of poverty...than to have defaced an enterprise of so great assurance...[11]

The only trouble was that no-one believed him. It was only later, when Ralegh became a hero of the anti-Stuart whigs, and subsequently of the Victorian imperialists, that his strategic genius came to be appreciated and his selfless nobility as a would-be empire builder was at last understood. Then he could be set alongside Clive of India and Cecil Rhodes as a model of virtuous ambition: aptly enough, I might add.

Privateering did not do much good for Roanoke.[12] It has been established beyond reasonable doubt that one of the main purposes of founding a colony there was to provide a base for privateering action against the Spanish treasure fleets bound from Havana to Seville and against the ports and shipping of Spain's Caribbean in general. My own view is that it was easily the main object, since I now believe that most of the talk about the value of a colony for the realm, in terms of social and economic benefits, was, if not exactly window-dressing, a form of argument, more or less persuasive, more or less sincere, offered to justify and gain support for a project which was not really designed to absorb excess population or provide markets for the cloth industry. But I have said what I think about that elsewhere.

In the first instance privateering appeared to be a very convenient way of meeting the expenses of the colonial enterprise. After depositing Ralph Lane and the first colony on Roanoke in 1585, Sir Richard Grenville captured a rich Spanish prize on his return voyage, which may well have balanced the books for that year. But in the following year the supply ship promised by Ralegh for the relief of the colony did not arrive until late June, by which time the colony had been abandoned. We don't know why the relief was so late, but we do know Ralegh had at least two ships out prize-hunting that summer. Meanwhile Grenville sailed for Roanoke with a squadron of West Country privateers, but delayed and depleted the expedition on the way out by taking and manning home various prizes, so that when he reached Roanoke in mid-July he found the place deserted, and made the unfortunate decision to leave a small holding party. He probably could have left a more substantial force had he not wished to cruise for further spoil on the return voyage. So he did in fact take more prizes, but failed to re-establish the colony: the fifteen men he left there were never seen again.

Privateering therefore was already proving an unfortunate distraction from the colonial enterprise. It held out immediate and by no means illusory prospects of large-scale loot, whereas colonization seemed to be simply a bottomless bag into which money poured and disappeared. And the second colony under Governor John White suffered even more from the competition of the plunder business. Roanoke having proved unsuitable as a settlement site, it had been decided to place the new colony in Chesapeake Bay. The ships went first to Roanoke, however, to contact the men left by Grenville, and when they arrived the sailors, led by Simon Fernandes, insisted on leaving the planters there, on the grounds that 'the summer was far spent', meaning that they were not prepared to waste any more time settling White's people when there was more lucrative work to be done at sea. This was the first instance of many in which seamen treated settlers with scant regard, but in this case they had a clear motive. Then in 1588, White—having returned to England—managed to get two pinnaces to sea with supplies for the stranded colonists, but failed to make Roanoke because the pinnaces took to chasing prizes and were treated to a dose of their own medicine, returning to port crippled. The next year nothing was done, though numerous men of war were cruising the Atlantic and it was a great year for prizes, of which Ralegh had his share. Finally in 1590 White shipped with a privateering squadron of which two vessels found time to visit the Carolina Banks and found traces of the lost colonists. After this Ralegh took to privateering with a will and pursued a course of plunder throughout the rest of the war and Elizabeth's reign, at the end of which he is to be found pleading for the continuation of the conflict, above all at sea, Spain's shipping and colonies being then, he argued, ripe for a last and devastating campaign of 'pickery'.

It was only after the war and in the seclusion of the Tower that Ralegh laid the foundations of his reputation as an authority on naval warfare, maritime strategy, the expansion of England at Spain's expense and the new rivalry between this country and Holland. Ralegh's literary output on these themes deserves perhaps more careful analysis than it has received to date, but it is clear at least that he succeeded in presenting himself as the champion of militant maritime expansion. This courtier, the favourite who owed everything to Elizabeth, now, after her death, criticised her timid conduct of the war, suggesting that a bolder policy of aggression would have reduced the king of Spain to a king of figs and oranges. 'Her majesty', he pontificated, 'did all by halves, and by petty invasions taught the Spaniard how to defend himself, and to see his own weakness.' Now is not the time to discuss the justice of this attack, but it is relevant to notice that it came from the most inveterate exponent of petty enterprise, who personally promoted strategically valueless commerce-raiding throughout the war and was totally committed to the type of naval organization which ruined Elizabethan offensive operations over and over again. That was the joint-stock naval enterprise, financed by the crown and syndicates of private subscribers investing for profit: in fact, privateering writ large. Such were the great expeditions of 1592, 1596 and 1597, in each of which Ralegh played a major role.

There was always something bogus about Ralegh, and about his reputation as a naval commander, a strategist, an imperial pioneer. I say nothing of the sanctimonious manner he chose to adopt in his *History of the World*, in contrast with his former scepticism. He was, as Greenblatt maintains in his biography, a great performer, who continually as it were staged his own life, and finally his own death.[13] We should be wary of accepting such a man on his own terms, judging him by what he wrote or was reported as saying. His personality was and is dangerously deceptive; even Christopher Hill has adopted him as some sort of proto-bourgeois revolutionary, whose heart was in the right place.

Meanwhile what of Elizabethan privateering, its character and its historical significance?[14] The most obvious feature to the researcher is its peculiar legal status. In fact it was, ironically enough, grounded in legalism. Privateers were vessels equipped with letters of reprisal, which were issued by the High Court of Admiralty and declared the legally-established right of the holders to recover damages for losses specified in terms of pounds sterling. The losses were supposed to be proved by due process of the court and indeed in the initial stages of the war such proceedings did take place on behalf of merchants whose property had been confiscated in Spain by the embargo of May 1585, which was thus the occasion for the outbreak of maritime hostilities. Before long, however, as the *de facto* state of war became obvious, though not recognized *de jure*, the procedure became a formality by means of which those who wished to set forth expeditions to plunder Spanish shipping could, in effect, buy the requisite letters of reprisal from the court—which meant in reality from the Lord High Admiral, Charles Howard of Effingham, later earl of Nottingham. Letters of reprisal would also frequently be bought by shipowners for trading voyages, so that should the occasion arise a merchantman on a Mediterranean voyage, for example, could bring home a prize. Licences were thus made available to all and sundry at a price. The court could of course refuse, but I have never come across such a refusal, while on the other hand the Lord Admiral did not object too strongly if a prize was taken without letters of reprisal, provided he received the ten per cent of its value which was his due.

The official licensing of privateers by this system was intended as a measure of control. Promoters and captains of each expedition had

13

to give bonds to observe certain rules laid down by the admiral: to confine their attentions to Spanish ships and goods, to bring prizes back to their home port, not to break bulk (that is, to rifle the cargo before it was inventoried and appraised by the admiralty officers), and so forth. And of course it was in the interests of the Lord High Admiral and his officials that this system should be enforced, since it was in effect a system of taxation. And there was the rub. Everyone knew that the Admiralty was a private empire, the proceeds of which sustained Howard's great personal wealth and power and the vested interest of his vice-admirals, deputy vice-admirals, tenth-collectors, secretaries, judges, sergeants, his family, his clients, his favourite captains and so forth—a large and prosperous following, many of whom were themselves promoters of ships of reprisal, investors or participants in the great joint-stock enterprises of the sea war, or wholesale dealers in prize-goods.

The Admiralty was the outstanding example of that typically Elizabethan compromise between the government and the vested interests, by which the former nominally administered the latter but the latter in reality manipulated the former. It was a perfect recipe for corruption. Privateering was bound in any case to be a somewhat disorderly business, requiring a strong hand for anything like effective control. Howard's was anything but a strong hand: a greedy one perhaps, but not effectively rapacious. A complaisant individual himself, evidently one who liked to be liked, he was inhibited by the fact that he himself was the greatest of the privateering magnates. Apart from being chief of the queen's navy, he had ships of his own. He sold at least one to the queen: that was the *Lion's Whelp*, a fast and strong pinnace of some fifty tons when he employed her as a privateer; but having been rebuilt, she performed poorly, and it was then he found a place for her in the royal navy. On another occasion he had one of his own ships commissioned as a royal vessel in an official expedition, so that her fitting out was done at the queen's expense, though Nottingham and his partners shared the profits. One of those partners was Sir Robert Cecil, then the queen's secretary, who with Nottingham promoted a series of expeditions into the Mediterranean for the dual purpose, as they stated, of 'reprisal of pirates and Spaniards'.[15] That is, the object was to arrest English pirates who were causing mayhem and much diplomatic embarrass-

14

ment there, whilst at the same time picking up Spanish prizes to recompense the entrepreneurs and their captains and crews. As might have been expected, however, these were none too scrupulous themselves about plundering Italian and French shipping and goods and disposing of them in Algiers. Thus Cecil and Nottingham succeeded in increasing the menace of English piracy in the Mediterranean and in forging the first links between English pirates and the Barbary corsairs. The evidence suggests that they did manage to compensate themselves financially for these misfortunes, but they had to arrange some complicated legal manoeuvres to achieve this end: with the necessary assistance of that illustriously named judge of the Admiralty Court, Sir Julius Caesar.

As I have said, privateering was inevitably a disorderly business, and under Howard's aegis it became very disorderly indeed. Not that this was entirely—nor even I think principally—his fault. In the first place the queen's navy was too small to undertake wide-ranging or regular operations against the enemy's fleets and shipping. This was traditionally the task of voluntary ships: some of them powerful enough to be considered men-of-war, most of them representative of the mass of the nation's shipping. Reliance on these to damage Spanish commerce was inevitable, and since Spanish commerce was rich, tempting and vulnerable there were plenty of volunteers. I have calculated that private ships engaged in reprisals numbered at least a hundred a year, and at times more like two hundred, and I now feel that my original estimates were on the modest side. The middling and small privateers who formed the majority carried no great armament, relying rather on minor artillery and men armed for boarding with calivers, swords, daggers and such hand weapons. For it was manpower that mattered in this business, and the readiness of owners to cram their ships with men was only exceeded by the eagerness of seamen to serve in such ventures. Manpower was cheap; ports were swollen with men at a time of growing population pressure and the nation's trades were disrupted by the war itself. Privateering not only absorbed the numerous pirates of the pre-war period, inheriting their attitudes and practices, but it drew upon the whole maritime population, inducting landsmen as well as seamen *en masse* into a kind of predatory voyaging which verged upon and frequently deteriorated into piracy.

15

For the most part promoters did not pay wages, but engaged men who were prepared to sail for thirds and pillage. A third was the customary share of the prize cargo allotted to the privateer crew as a whole, to be divided among them according to rank. The other two thirds were allotted to the owners and victuallers. Pillage was the customary right of the crew to whatever did not belong to the prize cargo proper: for example, all goods found above deck, the possessions of the prisoners, various items of equipment such as cables, and so forth. Pillage of a prize took place upon its capture, of course, and was supposed to be carried out in an orderly manner, the relevant goodies being assembled and divided according to customary rules. Custom, however, was always a matter for argument, and argument would tend to get heated, particularly when the prize had wine aboard, or best of all had a cargo of wines. We hear of 'brawls that were on board amongst the company' in one of Sir George Carey's privateers, the men having 'drunk that day well of the wines which they found in the prize, and some more than did them good, and that made them lie together by the ears the most part of the day'.[16] Pillage was very often extended to the cargo itself, which was rifled unmercifully if it consisted of valuables the sailors could conveniently stow away. The case of the great Portuguese carrack the *Madre de Deus*, captured in 1592 by a number of privateers in consortship, is notorious: the whole night after her capture was spent in an orgy of pillaging money, jewels, gold chains, silk, ambergris and so on, so that when she was brought into Dartmouth the crews from the prize and her escorts staggered ashore laden with wealth of which they were speedily relieved by the businessmen of the town and those who had already raced thither from London (and, no doubt, Exeter) to share the spoil. Commissioners were sent down by the government, and Ralegh, who had disgraced himself with Bess Throckmorton, was released from gaol to use his influence with the mariners, but of this, certainly the richest prize of the war, only a fraction of the whole in value eventually came to the promoters.

Apart from pillage, privateering crews were frequently accused of embezzlement: that is, disposing of the cargo before the owners and victuallers could secure their shares and cheating the Lord Admiral and the queen's customs of theirs as well. One can hardly blame the sailors for this sort of thing; if they behaved themselves like good boys they would certainly be cheated by the shipowners, merchants and gentlemen who were not prepared to pay them wages in the first place. Nor do I blame them for indulging in piratical attacks upon neutral shipping, even when such ships were not colouring Spanish goods. In the circumstances of the time it was every man for himself and the devil take the hindmost. What I object to is the tendency of later generations to treat these men as heroes, fighting for some higher cause than their own survival in a hostile world. Of course they were patriotic, in the sense that they regarded one Englishman as better than two froggies or six dagoes, but people should look to the records rather than to Kingsley or Newbolt to taste Elizabethan reality raw, not cooked in syrup. Of course they often appear in the records as humble and obedient servants, but the mutinousness of Elizabethan crews at sea and their drunkenness ashore cannot be ignored. They were not well regarded, as Captain John Smith's words convey: 'however slightly soever many esteem sailors', he said, 'we depend on them for the safety of ship, goods and lives.'[17] Smith stood up for them, but the general public had a pretty low opinion. To contemporaries they did not seem to embody the virtues of the sober, industrious and god-fearing Englishman.

As a means of waging war privateering was ineffective, and in so far as the privateering interests came to dominate the conduct of the war at sea in the semi-official joint-stock naval operations, its effect was to weaken the striking power of the state, ruining particular expeditions and generally holding back the development of the navy as a strategic weapon. It probably helped nevertheless to undermine Spain's merchant marine and in some small measure aggravate that country's economic difficulties. It also assisted the accumulation of wealth, shipping and experience by groups of merchants, gentlemen and captains who were to play a leading part in the Jacobean efforts to colonize parts of North America and develop direct trade with the East. These were considerable contributions to English expansion overseas, but they were in no sense planned or deliberate. The men of the privateers, whether promoters or seafarers, were not doing their bit towards a future British Empire.

There were, however, some people trying to romanticise them already, so that the beginnings of myth may be found in some of the writings of the day. We have noticed Churchyard's verse, and there was still worse doggerel by Henry Roberts the privateer-poet:

Brave noble brutes, ye Trojan youthful wights,
Whose laud doth reach the centre of the sun:
Your brave attempts by land, on seas your fights,
Your forward hearts immortal fame hath won....[18]

This was the opening of a broadside relating a fairly ordinary privateering cruise by one Grafton, described by Roberts with great enthusiasm as a Londoner, a citizen and not a cavalier.

Another Roberts product was 'The trumpet of fame', celebrating Sir Francis Drake's departure for the West Indies on his last voyage. And the same voyage inspired the excruciating work by Charles Fitzgeffrey, 'Sir Francis Drake his honorable life's commendation and his tragical death's lamentation', which can only be compared with Pyramus and Thisbe in *Midsummer Night's Dream*.[19] The web of myth was already being spun in Elizabethan England, most diligently by Dr John Dee, the wizard of Mortlake, alchemist, spiritualist, mathematician and cosmographer. He was also the first inventor of the phrase 'British Empire', which in his vocabulary had a decidedly Welsh flavour, incorporating for example the legend of Prince Madoc, who allegedly began the colonization of Dee's *Atlantis* (viz America) in the twelfth century, and so, we may infer, paved the way for Gilbert and Ralegh—at least in the realm of imagination, which was by no means unimportant. But more seriously the foundations of myth were laid in the early Stuart period, when men were looking back to the glorious days of the late queen and the victorious war with Spain. Ralegh had a considerable say in this, for although he criticised Elizabeth for not being as victorious as she should have been, he also contrived to suggest that she was better than James, who was letting the navy rot.

Perhaps more representative of common opinion was Robert Johnson's *Nova Britannia*, issued in 1609 as part of the desperate campaign to drum up support for the Virginia colony, which was on the verge of collapse:

It is known to the world and cannot be forgotten that the days and reign of Queen Elizabeth brought forth the highest degree of wealth, happiness and honour that ever England had before her time...I do only call to mind our royal fleets and merchant ships (the jewels of our land), our excellent navigators, and admirable voyages, as into all parts and round about the globe with good success, to the high fame and glory of our nation; so especially their aim and course was most directed to the new found world, to the main land and infinite islands of the West Indies, intending to discover with what convenience to plant and settle English colonies; wherein after many tedious and perilous adventures, howsoever strange seas and miserable famine, [etc. etc.]...daily armed with invincible courage and greater resolution (scorning to sit down by their losses) made new attempts....[20]

Here Hakluyt's influence is almost certainly at work. And so we find that the sordid reality is already being transmuted into the romantic image, which was not after all an invention of Kingsley's however much he contributed.

Analysis of the myth and its making requires a lot more thought and research than I can pretend to have given it. Myth, propaganda and illusion had as important a part to play in the creation of the empire as in its expansion and glorification in Victorian and Edwardian days. It would be interesting to trace particular themes through—the religious justification, for example, or the argument relating imperial expansion to social welfare. Such ideas have been identified and studied in their nineteenth century context. It is seldom realised that analogous ideas are to be found in Elizabethan times, though of course the ideology as a whole and in particular respects was different. Nor should we assume that it was generally accepted, any more than the imperialist ideas of Chamberlain, Seeley, Kipling and company were in their time. Inquiry along such lines could, I believe, do something as an antidote to counteract the tendency to sentimentalize Elizabethan maritime expansion and the origins of the British Empire, a tendency which seems to me to be gaining ground now more than ever. Precisely why this should be so I fail to understand, but nostalgia is an insidious and contagious disease: we would do well to build up resistance to it.

NOTES

1. Kingsley's historical interests and attitudes strongly resembled those of his close friend and brother-in-law, James Anthony Froude. Both were disciples of Carlyle and it was the controversy over Kingsley's review of Froude's *History of England*, vols vii and viii, which provided the occasion for J. H. Newman's 'Apologia'.

2. For the location of these statements see K. R. Andrews, *Trade, Plunder and Settlement: Maritime Enterprise and the Genesis of the British Empire* (Cambridge, 1984), p. 28.
3. G. E. Manwaring and W. G. Perrin (eds), *The Life and Works of Sir Henry Mainwaring* (2 vols, London, 1922), II, p. 18.
4. E. S. Donno (ed.), *An Elizabethan in 1582: the Diary of Richard Madox* (Hakluyt Society, 2nd Series, vol. 147: London, 1976), p. 144.
5. J. K. Laughton (ed.), *State Papers relating to the Defeat of the Spanish Armada* (2 vols, London, 1895), II, p. 102.
6. D. B. Quinn (ed.), *The Voyages and Colonising Enterprises of Sir Humphrey Gilbert* (Hakluyt Society, 2nd Series, vols 83–84), p. 339.
7. *Ibid.*, p. 217.
8. V. T. Harlow (ed.), *The Discoverie of Guiana by Sir Walter Ralegh* (London, 1928), pp. xxvi–xxix.
9. *Ibid.*, p. 4.
10. Cited in K. R. Andrews, 'Sir Robert Cecil and Mediterranean Plunder', *English Historical Review*, 87 (1972), 513.
11. Harlow, *op. cit.*, p. 6.
12. The following analysis of the contribution of privateering to the Roanoke enterprise is based on D. B. Quinn (ed.), *The Roanoke Voyages, 1584–1590* (Hakluyt Society, 2nd Series, vols 104–5: London, 1955).
13. S. J. Greenblatt, *Sir Walter Ralegh: The Renaissance Man and his Roles* (1973).
14. See K. R. Andrews, *Elizabeth Privateering: English Privateering during the Spanish War, 1585–1603* (Cambridge, 1964).
15. See Andrews, 'Sir Robert Cecil and Mediterranean Plunder', *loc. cit.*
16. K. R. Andrews, *English Privateering Voyages to the West Indies, 1588–1595* (Hakluyt Society, 2nd Series, vol. III: Cambridge, 1959), p. 151.
17. John Smith, *Works*, ed. E. Arber (2 vols, 1884) II, p. 803.
18. Andrews, *Elizabethan Privateering*, p. 150.
19. K. R. Andrews (ed.), *The Last Voyage of Drake and Hawkins* (Hakluyt Society, 2nd Series, vol. 142: Cambridge, 1972), p. 3.
20. D. B. Quinn (ed.), *New American World: a Documentary History of North America to 1612* (5 vols, London, 1979), V, p. 236.

2

The Three-masted Ship and Atlantic Voyages

Ian Friel

The term 'three-masted ship' is a generic one, used to describe a vessel of the fifteenth or the sixteenth centuries which combined a particular form of hull construction with a particular form of rig. The three-masted, or full-rigged ship, as it is sometimes known, played a crucial role in the early European oceanic voyages. This paper examines the origins and development of the type in the Middle Ages and discusses the shipping used by the English in their Atlantic ventures up to the time of the Roanoke voyages. This discussion has to remain on a fairly general level, because very little is known about the majority of these vessels. For example, the accounts of the Roanoke voyages do not even name the two ships used for the first expedition in 1584.[1] The three-masted ship was a product of the amalgamation of different traditions of medieval maritime technology, and in order to understand how it originated we have to look first at the ships of Europe in the later Middle Ages.

There were two major and distinct systems of ship construction and rig in medieval Europe. North European ships were generally shell-built in the clinker fashion (also known nowadays in the United States as 'lapstrake'): this means that the hull was formed from a shell

21

of overlapping planks. The planks were fastened together at the edges by nails whose points were clenched over iron washers called 'roves' on the inboard faces of the planks (the clenching process giving rise to the term 'clinker'). Frames were not inserted into the hull until construction of the plank shell was reasonably well-advanced, to give the hull form some rigidity (see Fig. 1). In the final structure it was the shell that was the main load-bearing element, not the frames.[2] The clinker-built ship could be strong, durable and flexible, and as the Viking voyages to Greenland and beyond showed, it was more than up to the rigours of oceanic voyaging. However, there were drawbacks to the clench-fastened hull: it could be difficult to repair, and it was relatively expensive in terms of materials and skilled manpower. All of the clench-nails needed separate metal components in the form of roves, and the production of the hull shape required the work of a comparatively large number of the more skilled, and therefore more expensive, shipwrights.[3]

Shipbuilding techniques in the Mediterranean and Southern Europe were a complete contrast to those in the North. Construction of a Southern hull began with the erection of a skeleton of frames

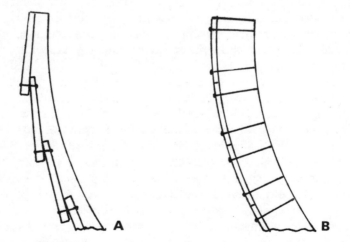

Fig. 1. Schematic sections through the sides of shell and skeleton-built hulls. A: Clinker construction. Note nails fastening planks at edges. B: Skeleton construction. Note nails fastening planks to frames.

onto which flush-laid planks were subsequently nailed, giving the hull exterior a smooth surface. This method is known as skeleton construction, and it developed in the Mediterranean at some point between the sixth and eleventh centuries AD, superseding an earlier form of non-clinker shell construction. With skeleton construction the main strength of the hull lay in the skeleton, not the shell of thin planking, and the most difficult part of the building process was in the shaping and setting of the frames. Actually nailing on the planking was still time-consuming and labour-intensive, but it did not have to be done by master-craftsmen. Skeleton construction needed less wood than ancient Mediterranean shell construction, and although the nails used could be large, they did not require roves as did Northern clench-nails. The skeleton-building technique was cheaper to use than either of the two shell-building methods, and it has been suggested that its development was prompted by the shortages of materials and skilled workmen in the disturbed times following the collapse of the western Roman Empire in the fifth century AD.[4]

Ship rig in medieval Northern and Southern Europe also differed very markedly. Northern ships were square-rigged, with a single square or rectangular sail on a mast set amidships. The medieval square sail was very useful when the wind was behind or to the side of a ship, but it was not always well adapted for sailing into the wind. Modern trials with replicas of Viking ships have indicated that such vessels could sail as much as 30° into the wind in favourable conditions: modern yachts are designed to sail only some 15° closer to the wind. However, the fact that this can be done with reconstructions does not mean that sailing close-hauled (i.e. close to the wind) was commonly attempted in the Middle Ages or could be so successfully achieved with the many vessels that were not so well-designed or well-built as those of the Vikings.[5]

The square sail was known in the Roman Mediterranean, but in the early Middle Ages it was supplanted by the triangular lateen sail. The lateen is rigged along the fore-and-aft line of a ship, and like the fore-and-aft rig of a modern yacht it is specifically designed for sailing to windward: a lateen-rigged ship can sail as much as 30°–35° into the wind, but it is less efficient in a following wind and can prove unstable.[6] In bad weather the medieval square sail could be reefed in,

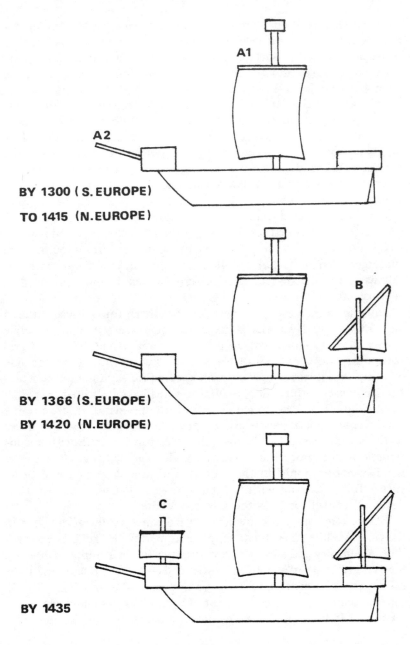

A1

A2

BY 1300 (S. EUROPE)
TO 1415 (N. EUROPE)

B

BY 1366 (S. EUROPE)
BY 1420 (N. EUROPE)

C

BY 1435

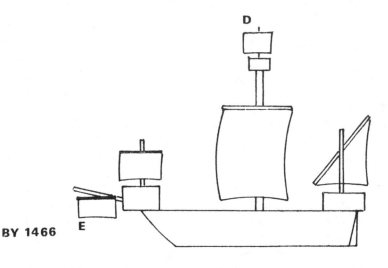

BY 1466

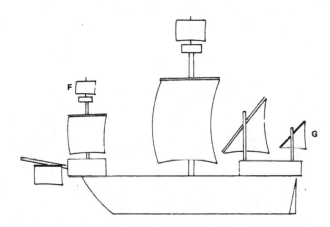

BY 1500

Fig. 2. The development of the rig in Europe, c. 1300–1500. **A1**: square mainsail and yard. **A2**: bowsprit. **B**: lateen-rigged mizzenmast. **C**: square-rigged foremast. **D**: square-rigged topmast. **E**: square spritsail and sprityard. **F**: fore topmast. **G**: after-mizzen or bonaventure-mizzen. *Note*: diagrams are schematic, and the standing and running rigging have been omitted. Sources for Fig. 2: see notes 5–15.

or the sail area could otherwise be reduced by removing detachable portions of canvas called 'bonnets' from the bottom of the sail. The only way to reduce sail area on a lateen-rigged vessel in the Middle Ages was to remove one sail and put another in its place. Tacking a lateener (that is, going across the wind to bring it on to the other side of the ship) involved raising the sail and yard vertical and walking them round to the other side of the mast. On a square-rigged ship, all that needed to be done was to pull the yard and sail round, using the running rigging. Lateen rig was more cumbersome and labour-intensive than square rig, and therefore more costly.[7]

The cost advantage of square rig probably lay behind its adoption in the Mediterranean in the late thirteenth or early fourteenth centuries. Southern shipwrights and mariners adapted the hull form and single square rig of the Northern cog to their own use, producing a skeleton-built bulk carrier which they called a *cocha* and was later known in the North as the carrack. The hull shape of the cog may have been cheaper to build and more efficient for carrying bulk goods than those of contemporary Mediterranean types, and its rig could be operated by fewer men than the lateen sail required. *Cochas* of many hundreds of tons burden (cargo capacity) were produced, but the great size of these vessels meant that the wind pressure on the hull and superstructures could be considerable, creating sailing problems. A step towards solving these problems was taken in about the middle of the fourteenth century when a second, smaller mast was added to the *cocha*, stepped behind the mainmast. This was called the mizzenmast, and the mizzen sail was a lateen whose good windward sailing qualities helped to improve the manoeuvrability of the *cocha*[8] (for this and subsequent rigging developments, see Fig. 2).

The two-masted square rig (so-called despite the presence of the lateen) does not seem to have been used in the North until the early fifteenth century. The naval forces of Henry V of England (1413–1422) captured a number of two-masted Genoese carracks, and there is evidence that attempts were made to rig some of the larger or longer royal ships as two-masters. Like the carracks, their sailing qualities could have been improved by this rig. However, the two-masted rig was a relatively short-lived phenomenon on the way to the development of the three-masted square rig.

The earliest-known documentary evidence for this form of rig comes from England. In 1420 Henry V's massive 1400-ton 'great ship' *Grace Dieu* had three masts and sails, although unfortunately we do not know where the third mast was stepped. It is possible that this mast was a foremast, carrying a small square foresail: a sail of this kind was certainly part of the equipment of an oared royal balinger, the *Petit Jesus*, built in 1435. The rig of the *Petit Jesus* comprised a mainsail, mizzensail and *fokesail* or foresail, the basic components of the three-masted square rig. The square foresail was used to supplement the work of the mainsail in propelling the ship, but it also served as a manoeuvring sail, helping to balance out the effects of the mizzen. The foresail also provided a more effective means of bringing the ship's head round on to a more favourable tack. We do not know if the foresail and foremast were English inventions, although the great size of the *Grace Dieu* (she was probably over 200 ft in length) may have prompted some experimentation with her rig in order to improve her sailing and steering. Whatever the antecedents of the foresail, it is clear that the three-masted square rig was known in Northern Europe by the 1420s or mid-1430s.[9]

The transmission of skeleton construction to Northern Europe seems to have slightly post-dated the development of the three-masted rig. The details are rather difficult to unravel, but the initial agents in the spread of this technique seem to have been the Portuguese, who from the late 1430s were using caravels on their voyages to Northern Europe. The caravel was a small, lateen-rigged ship with a skeleton-built hull, a vessel of major importance in the early Portuguese voyages of exploration. North Europeans called them carvels, but at first they were unable to build them because the skeleton construction technique was alien to their shipbuilding traditions. It is not until the 1450s and 1460s that we begin to find appreciable numbers of carvels in North European ownership:[10] at about the same time there are indications that Northerners were learning how to build them, either by studying carvel hulls or by being taught the skeleton technique by itinerant shipwrights. It is no coincidence that in modern English, Dutch, German and other North European languages the term 'carvel' is synonymous with skeleton construction. By the late fifteenth century knowledge of this shipbuilding method had spread throughout Northern Europe.[11]

Skeleton construction ousted clinker building as the main method for building ships in the North. Although the technique survives to this day as a way of making small boats, clinker construction died out as a means of building large ships in the early sixteenth century. The reasons for this change were probably both economic and technical. Skeleton construction required fewer skilled shipwrights than clinker building, and was less expensive in ironwork for nails and probably also in timber for making the planks. With their large frames, skeleton-built hulls were perhaps stronger than clench-fastened ones and the flush-laid planks were easier to maintain and repair than the overlapping clinker planking.[12]

By the middle of the fifteenth century the three-masted square rig and the skeleton-built hull had been combined. The earliest evidence for this is to be found in the equipment of a Spanish caravel of 1453, which had a mainmast, mizzenmast and foremast.[13] The combination of the new rig with the skeleton-built hull represented the final merging of the two medieval shipbuilding traditions, producing a type of vessel that was distinctly post-medieval.

The earliest-known carvel to have been built in England was also a three-master. This 'new kervelle' was constructed in Suffolk for Sir John Howard between 1463 and 1466. The building account shows that the rig of this ship incorporated two new developments besides the basic three masts. A yard was set under the bowsprit to carry a square sail called a spritsail: this provided extra canvas ahead of the ship's bows to help with propulsion and manoeuvring, and had a significant effect in increasing control over the vessel. The carvel also had a topmast, set in the topcastle at the head of the mainmast. This was to carry a small square sail which could catch the more constant breezes at masthead height and tended to lift the bows of the ship, counteracting the spritsail's tendency to bury them in the waves. In essence, the rig of Sir John Howard's carvel was little different from that of Columbus's *Santa Maria* of 1492.[14]

In the late fifteenth and early sixteenth centuries the division of sail-plan increased. By the mid-1480s some ships had a second, lateen-rigged mizzenmast called the after-mizzen or bonaventure mizzen, stepped behind the larger mizzenmast. Topmasts appeared on the foremast and mizzenmast, and on some of the larger and more prestigious warships and trading carracks we begin to find square-rigged topgallant masts above the topmasts on the fore, main and mizzenmasts.[15] However, the majority of oceanic sailing vessels of the sixteenth century were three-masters with rig little changed in its essential features from that of Howard's carvel. The only major additions were the fore topmast and the mizzen topmast, and even this latter development was not always used.

In terms of medieval technology, the development of the three-masted ship was very rapid indeed. Little more than a century after the appearance of the two-masted square-rigger in the Mediterranean the two medieval shipbuilding traditions had combined to produce a type of vessel that had five or more sails and a strong, skeleton-built hull, used in both Southern and Northern Europe. The three-masted ship became the mainstay of the European voyages of exploration and colonisation, but I do not believe that it was produced with these ends in view, because it had appeared before anyone but the Vikings and the Portuguese had made serious efforts in oceanic sailing. The new type evolved simply because it was a better kind of ship than anything that had preceded it. The three-master was more manoeuvrable than the old single-masted square-riggers or the lateen-riggers. Early-seventeenth century authorities said that a square-rigged ship of their day could sail up to $32\frac{1}{2}°$ to windward, and closer if conditions were favourable.[16] Whilst this windward performance was not quite as good as that of the lateen-rigger, the three-master could tack much more easily and could put on a great deal more canvas in a following wind. This made the new type especially useful in sailing the prevailing oceanic winds, such as the North-East Trades in the Atlantic. The greater divisibility of sail-plan also increased the number of options open to a mariner in deciding how to rig the sails to meet the very variable weather conditions found on the oceans, thus increasing the ship's chances of survival. The crewing costs of the three-master were not significantly greater than those of its predecessors because each sail could be rigged in turn. The rig, the strong and durable hull and the economic crewing levels of the three-master helped to make it a viable proposition for oceanic voyaging. It was the right development at the right time.

Another innovation associated with the three-masted ship revolution was the invention of the gunport. The strength of the skeleton-built hull meant that numerous ports could be cut in it to enable guns to be carried lower down in the hull. This allowed ships to carry much heavier artillery, and led to one or more of the lower decks being used exclusively as gundecks. Gunports first appeared at the beginning of the sixteenth century, and marked the start of a division between sailing warships and merchantmen, a process not complete until the seventeenth century or later. Although merchant ships continued to carry armament, they could not carry as many guns as the purpose-built warships because the gundeck space encroached on that available for cargo. However, as we shall see, warships played a not inconsiderable part in the early English overseas ventures.[17]

To understand the types of ships available for the English oceanic voyages between the 1480s and 1580s we have to know something about English merchant shipping in this period. Table 1 summarises

Table 1: the tonnages of English merchant shipping, 1449–1582 (tonnage percentages rounded to nearest whole numbers)

Date	Type of record	Total no. of ships	Tonnages:			
			0–99	100–99	200–99	300 +
1449–50	A	161	83	44	27	7
		(100%)	(52%)	(27%)	(17%)	(4%)
1461–67 and 1472–83	C	61	23	17	13	8
		(100%)	(38%)	(28%)	(21%)	(13%)
1512–14	A	167	110	45	7	5
		(100%)	(66%)	(27%)	(4%)	(3%)
1513	S*	85	56[1]	29	—	—
		(100%)	(66%)	(34%)		
1544	S*	240	207	33	—	—
		(100%)	(86%)	(14%)		
1560	S*	77	n.c.	70	5	2
		(100%)		(91%)	(6%)	(3%)
1577	S	791	656[2]	121	10	4
		(100%)	(83%)	(15%)	(1%)	(1%)
1582	S	1630	1453	158	16	3
		(100%)	(89%)	(10%)	(1%)	

A: ships arrested or hired by the Crown.
C: ships granted safe-conducts.
S: survey of ships;   S*: survey of part of coast only.
[1]ships of 60–99 tons only.
[2]ships of 40–99 tons only.
n.c.: not counted.
Sources: see note 18.

the (rather imperfect) statistical information available for the sizes of English merchant ships in this period. Despite the defects of the statistical data, a general trend is clear from the early sixteenth century: namely the decline in the proportion and numbers of vessels of more than 100 tons burden in the English merchant fleet and a consequent rise in the proportion of small vessels. It was not until the late 1570s or early 1580s that the actual number of ships of more than 100 tons burden began to appreciably increase, although they were still vastly outnumbered by the small ships. The rise in the number of large vessels was due to the growth in long-distance trade-routes to Africa and the Mediterranean.

Various reasons have been put forward for the 'ascendancy of the small ship'. Most English trade until the later Elizabethan period was of short range, to the Continent, and large vessels were too expensive to crew and maintain for such work. Small ships were better suited to medieval harbours and were comparatively cheap to operate. They also represented less of a risk for the shipowner's investment. However, this phenomenon was not just confined to England, but occurred all over Northern Europe and as far south as Spain.[19] Undoubtedly economic factors were also at work here, but it does raise the possibility that the 'rise' of the small ship was in part due to the adoption of skeleton construction in the North. Caravels were generally vessels of less than 100 tons, and in copying their construction it is unlikely that Northern shipwrights would have immediately started using the technique to build large ships. There is some evidence from England that the method of constructing large, skeleton-built warships was not mastered until the early sixteenth century.[20]

Small ships predominated in the early English voyages across the Atlantic. The earliest-known venture, that of John Lloyd in 1480, is said to have been undertaken in a ship of 80 tons. The *Matthew* used by John Cabot in 1497 for his voyage to Newfoundland had a crew of 18–20 men, and according to one source her burden was 50 tons.[21] Most of the large vessels used in sixteenth century voyages across the Atlantic seem to have been royal ships, such as the 160-ton *Mary Guildford*, a well-armed and well-equipped three-master which made a formidable journey to Newfoundland in 1527, returning via the West Indies. Forty years later we find the 700-ton royal warship *Jesus of Lubeck* taking part in one of John Hawkins's ventures, the one that ended in disaster at San Juan de Ulua.[22] However, the majority of English ships operating in the Atlantic in the sixteenth century were probably like those to be found participating in the Newfoundland fishery. Customs records of the 1580s and 1590s suggest that the size of vessels employed in the fishery ranged from 40 to 80 tons, with most ships being of 60 tons or less. A contemporary expert reckoned that vessels of 70 tons were the best for this work, crewed by a master and 24 men, only half of whom needed to be experienced seamen.[23]

The accounts of the Roanoke voyages tell us very little about the ships used: the most we can usually hope for is a name and a tonnage, and sometimes not even these details survive. A small number of type-names recur, the commonest being ship, bark and pinnace, terms that were sometimes interchangeable. For example, Raleigh's *Dorothy* was described as a small bark in one source and apparently as a pinnace in another document; the 170-ton *Bark Bonner* which evacuated the colonists in 1586 is called a ship.[24] Broadly speaking, the terms ship and bark seem to have been applied to three-masted ships of 50 tons or more: by contrast the pinnaces were the smaller sea-going vessels, generally in the 20–50 ton range (at least in the context of these voyages).

Pinnaces seem to have been three-masters, although the drawings of the 1585–86 Roanoke expedition show one or possibly two-masted vessels working close inshore. These are too large to be ships' boats, because they have fore- and aftercastles, and so are probably pinnaces. Pinnaces were valued because they were small, fast and manoeuvrable: two new pinnaces built for the 1585 voyage were said to have been 'adjoined [to the fleet] for speedie services'. Some of them could move under oars, and they were very effective in acting as scouting and landing craft amid the shoals and sandbanks of the American coast, ferrying stores and people to and from the larger ships anchored offshore. Some pinnaces were decked and carried four or five guns. The 30-ton bark or pinnace *Brave*, used on the abortive 1588 voyage, had some form of superstructure and was able to carry eleven settlers housed in cabins of some kind (probably constructed on deck: the cabins and their occupants getting in the way of the crew when they tried to fight out a boarding action).[25]

The ocean-going ships and barks of the Roanoke voyages were generally of some 50–200 tons burden. The 1585–86 drawings show them as three-masters, mostly equipped with topmasts on all three masts, and all have pronounced forecastles and aftercastles. The size of these vessels made them slower and less maneouvrable than the pinnaces and prevented them from getting in close to the American shore. However, they could carry many more men and guns than the pinnaces. The *Little John* of the 1590 voyage was of 100–120 tons burden and is said have had a crew of up to 100 men and an armament of 19 guns, whilst the 80-ton *Moonlight* had 40 men and 7 guns.[26] These vessels and the majority of the ships used on the Roanoke voyages were privately-owned, and little information about them has survived. The best-documented ships of this period are royal vessels, and as a result we know a little more about one of the ships of the 1585 voyage, the Queen's ship *Tiger*.

The *Tiger* was originally built as an oared fighting ship, a galleass, in 1546. She was rebuilt in 1570 and it appears that it was this vessel which went to Roanoke in 1585. Her burden is variously given as 140, 160 and 200 tons, and according to a naval document of 1585 she could carry up to 28 guns. For the Virginia voyage she seems to have had only some twelve or thirteen guns, for a report by the Spanish ambassador to London states that she had five pieces per side and two cannon in the bows. A drawing of the *Tiger* is to be found on the map of the 1580 siege of Smerwick Bay in Ireland (see Fig. 3) and shows her with some five gunports in the side and two in the stern (the ship being otherwise an unremarkable three-master). Some dimensions of a *Tiger* survive in a copy of a navy list of 1591, but unfortunately it is not certain that this is the same vessel that went on the Roanoke voyage, and it has been suggested that she was exchanged for a merchant ship after her return from America. The ship of 1591 had a keel length of 50 ft and a beam (i.e. width) of 23 ft: these are the proportions of a merchant ship, not a warship. Even in cases where a ship is apparently well-documented, there can sometimes be tantalizing gaps in our knowledge.[27]

The only other major type of vessel used on the Roanoke voyages was the fly-boat. One of the ships of the 1585 voyage, the 120–140 ton *Roebuck*, is described as a fly-boat in English sources. According to the Spanish ambassador both the *Roebuck* and the 100–120 ton

Fig. 3. A three-masted ship: the *Tiger* in 1580 (standing and running rigging omitted. After *Mariner's Mirror*, Vol. 51, 1965, pl. 15 and Public Record Office MPF 81).

Lion (which he called a fly-boat) were acquired from the Dutch. Another fly-boat, of unknown name and tonnage, was used in the 1587 voyage with the *Lion*. The English word 'fly-boat' was a corruption of the Dutch *vlieboot*, and later also applied to the Dutch *fluyt* which superseded the *vlieboot*. *Vliebooten* were broad-beamed vessels of shallow draught with square, built-up sterns. In Dutch service they were two-masted, each mast having a square sail and one mast also being rigged with a spritsail (a fore-and-aft rigged sail, not the sail under the bowsprit). Unfortunately we do not know if the fly-boats of the Roanoke voyages had this typical Dutch rig or a three-masted square rig, but the type was an excellent medium-size cargo carrier, and this probably accounts for its use in the Roanoke ventures.[28] The other craft of the Virginia voyages were not ocean-going vessels: they were small craft, such as ship's boats, shallops and even a four-oared Thames wherry. Boats were essential for inshore transportation and exploration, being able to go where even the shallow-draught pinnaces could not.[29]

We know the tonnages of only thirteen of the twenty-two or so ships that set out on voyages to Roanoke. Seven of these were pinnaces or barks of less than 50 tons, two were vessels in the 50–100 ton range, three were between 100 and 150 tons, and one was of 170 tons. Five pinnaces of unknown tonnage were also used, and none of these are likely to have been larger than 50 tons, suggesting that at least half the vessels used (not all of which actually reached Roanoke) were very small. Three other ships may have been in the 100–200 ton range, but in the Roanoke ventures, as in the contemporary English merchant fleet, it was the small ship that predominated.[30]

The three-masted ship became the standard sailing vessel for sixteenth century Atlantic voyages, and indeed for most other European oceanic enterprises. It was a sturdy, manoeuvrable vehicle that made the business of exploration and colonisation less hazardous than it might otherwise have been. It is worth remembering that, for all the dangers of the Roanoke voyages, only one ocean-going vessel, a pinnace, actually sank en route. Although some people were lost in small boat accidents, the major disasters of Raleigh's Virginia colonies took place on land.[31]

NOTES

1. D. B. Quinn (ed.), *The Roanoke Voyages 1584–1590*, Hakluyt Society Second Series No. CIV, 2 vols, London 1955, p. 91. Henceforth Quinn, *Voyages*.
2. See, for example, B. Greenhill, *Archaeology of the Boat*, London 1976, pp. 175–258; S. McGrail, *Rafts, Boats and Ships from Prehistoric Times to the Medieval Era*, London 1981, pp. 42–43.
3. G. J. Marcus, *Conquest of the North Atlantic*, Woodbridge, 1980, chaps 6–16, *passim*; H. I. Chapelle, *Boatbuilding*, London 1969, p. 441. As regards the time taken up by the work of the most skilled shipwrights, see for example the following percentages of master shipwright/highest paid shipwright time from medieval building accounts in the Public Record Office London, Exchequer Accounts Various (E101): 5/7 Ipswich galley 1295, 44%, Ipswich barge 1295, 52%; 5/2, Southampton galley 1295, 37%; 42/39, 'Great Boat' 1400, 50%; 43/6 *Godegrace* 1401, 39%.
4. R. W. Unger, *The Ship in the Medieval Economy 600–1600*, London 1980, pp. 104–105.

5. *How Did the Viking Ships Sail?* Educational Department, Viking Ship Museum, Roskilde, Denmark, n.d., n.p.; T. C. Gillmer, 'The capability of single square rig: A technical assessment', in S. McGrail (ed.), *Medieval Ships and Harbours of Northern Europe*, B. A. R. International Series No. 66, Oxford 1979, p. 179. Gillmer suggests that a theoretical capacity of sailing 35° into the wind could have been achieved in good conditions.

6. J. H. Parry, *The Discovery of the Sea*, London 1974, pp. 12–14.

7. See B. Landstrom, *The Ship*, London 1976, p. 83, for an illustration of tacking a small lateen-rigged vessel.

8. Unger, *op. cit.*, pp. 183–188; P. van der Merwe, '*Coche seu nave* and carracks in the later 14th century' in *International Congress of Maritime Museums 4th Conference Proceedings*, Paris 1983, pp. 122–9 (henceforth *ICMM*).

9. I. Friel, 'England and the advent of the three-masted ship', *ICMM*, pp. 130–8.

10. J. van Beylen, *Schepen van de Nederlanden*, Amsterdam 1970, pp. 7–8; Friel, *ICMM*, p. 134.

11. Greenhill, *op. cit.*, p. 292; Unger *op. cit.*, pp. 222–4.

12. A. McGowan, *Tiller and Whipstaff: The Development of the Sailing Ship 1400–1700*, London 1981, pp. 21–2.

13. A. G. Sanz (ed.), *Historia de la Marina Catalana*, Barcelona, 1977, p. 61.

14. Friel, *ICMM*, p. 134; McGowan, *op. cit.*, pp. 13–14, 16–17.

15. B. Sandahl, *Middle English Sea Terms*, Vol. II, Uppsala 1958, pp. 9, 20, 37, 79 and 110–12.

16. G. E. Manwaring and W. G. Perrin (eds), *The Life and Works of Sir Henry Mainwaring*, Vol. II, Navy Records Society, Vol. 56, London 1922, p. 102; K. Goell (ed.), *A Sea Grammar…1627*, London 1970, p. 50.

17. McGowan, *op. cit.*, pp. 20–1.

18. Sources for Table 1: D. Burwash, *English Merchant Shipping 1460–1540*, Newton Abbot 1969, pp. 179, 180, 183 and 185; G. V. Scammell, 'English merchant shipping at the end of the Middle Ages: some East Coast evidence', *Economic History Review*, Second Series, Vol. XIII, 1961, p. 332; M. M. Oppenheim, *A History of the Administration of the Royal Navy 1509–1660*, London 1896, pp. 171–7.

19. R. Davis, *The Rise of the English Shipping Industry in the Seventeenth and Eighteenth Centuries*, Newton Abbot 1972, pp. 6–7; Scammell, *op. cit.*, pp. 333–4.

20. W. Salisbury, 'The Woolwich Ship', *Mariner's Mirror*, Vol. 47, 1961, pp. 81–90; Oppenheim *op. cit.*, p. 54; Greenhill *op. cit.*, p. 292.

21. J. A. Williamson (ed.), *The Cabot Voyages and Bristol Discovery under Henry VII*, Hakluyt Society, Second Series, Vol. CXX, Cambridge 1962, pp. 187–8, 206, 209 and 213.

22. D. B. Quinn, *England and the Discovery of America 1481–1620*, London 1974, pp. 127, 174–81; J. A. Williamson (ed.), *The Voyages of the Cabots*

and the English Discovery of North America under Henry VII and Henry VIII, London 1929, pp. 101–111; McGowan, *op. cit.*, pp. 21–2.

23. G. T. Cell, *English Enterprise in Newfoundland 1577–1660*, Toronto and Buffalo 1969, pp. 3–4, 130 and 132.
24. Quinn, *Voyages*, pp. 91 (and n.6), 98, 179 (and n.4) and 292.
25. *Ibid.*, pp. 179 (and n.5), 190, 256, 293, 562–9, 657; Unger, *op. cit.*, p. 264; for the ship illustrations see D. B. Quinn, *Set Fair for Roanoke: Voyages and Colonies 1584–1606*, Chapel Hill and London 1985, p. 68.
26. Quinn, *Set Fair for Roanoke*, p. 68; Quinn, *Voyages*, pp. 580 and 662–3.
27. R. C. Anderson, *List of English Men-of-War 1509–1649*, Society for Nautical Research, Greenwich 1974, pp. 9 and 13; Oppenheim, *op. cit.*, pp. 120, 123 and 156; Quinn, *Voyages*, pp. 178–9 (and n.6), 205, 228 (and n.6), 728; T. Glasgow, 'The shape of the ships that defeated the Spanish Armada', *Mariner's Mirror*, Vol. 50, 1964, pp. 179–80, 182–4; R. C. Anderson, 'The sixteenth century *Tiger*', *Mariner's Mirror*, Vol. 51, 1965, p. 194 and pl. 15). See also T. Glasgow, 'H.M.S. Tyger', *North Carolina Historical Review*, Vol. 43, 1966, pp. 115–21.
28. Quinn, *Voyages*, pp. 178–9 (and n.1), 728 (and n.4): the ambassador refers to two flyboats, apparently the *Roebuck* and the *Lion*; for the third fly-boat, see p. 516; Unger *op. cit.*, p. 262.
29. Quinn, *Voyages*, pp. 190, 256 and 599.
30. Quinn, *Voyages*, *passim*.
31. Quinn, *Voyages*, pp. 175, 611–12.

3

Did Raleigh's England Need Colonies?

Joyce Youings

My task is to take a look at Raleigh's England, both nationally and locally, and to consider the need, especially in the 1580s, to plant its people, in greater or lesser number, on overseas territory, not forgetting, although there will not be time to develop this theme, that, in David Quinn's words, Ireland was 'the moving frontier' of Elizabethan expansion overseas. I do not propose to discuss why, in the short run, Raleigh's colonising project failed, though some of the reasons will be implicit in what I have to say. Nor indeed shall I discuss Raleigh's own intentions and expectations. Rather I want to consider how far he, and those most closely associated with him, correctly judged the needs of Elizabethan England.

The mounting of the two main expeditions, those of 1585 and 1587, involved very considerable expense and, even allowing for the chance of prizes, great financial risk. It also involved the finding of men, and in 1587 of women, colonists and not least the engagement of crews to man the ships. But most difficult of all Raleigh had to persuade the Queen to allow him to proceed. To these ends, and especially the last, he needed intellectual support, the writing of promotional literature. The best known of his publicists were, of course, the two Richard Hakluyts, the elder a lawyer and the younger, his cousin, a clergyman, both essentially Londoners. Most of their extensive writings were eventually published and are now available in good modern editions, but in the mid 1580s they were confidential documents, circulated in manuscript.[1]

But not all who argued the case for colonies were armchair adventurers: some were would-be participants such as Christopher Carleill, Walsingham's stepson, who compiled his 'Discourse' in 1583. From nearly a decade earlier we have the splendidly-written appeal to the Queen, actually directed to her Lord High Admiral, the earl of Lincoln, of a group of west country gentlemen led by Sir Richard Grenville.[2] Most were his cousins, but also included was a certain Martin Dare. At this time, of course, Walter Raleigh was barely 20 years' old, and, having already been in France fighting with the Huguenots, was continuing his education at Oxford. Grenville's objectives in 1574 lay in the southern hemisphere, but he certainly envisaged the ultimate planting of settlements.

Running through these and other contemporary proposals certain common themes emerge, not the least interesting of which are those which link overseas colonisation with contemporary English society, themes such as population, employment, navigation, trade, religion and the idea of dominion, that is lordship, over land. Each of these is, of course, a subject in itself and some will have to be dealt with very briefly.

To start with population, Grenville's remarks in 1574 are interesting. He did not press the need to export people, though he did contend that the country contained 'more than can be provided for'. According to the calculations of Cambridge demographers, the country's population had by the early 1570s barely recovered from the effects of the devastating influenza epidemic which had preceded the Queen's accession, and had been further hit by outbreaks of bubonic plague of such dimension that London in 1563 lost some 20,000 people, nearly a quarter of her total population. Crediton in Devon in 1571 lost about the same proportion. The 1570s brought fewer epidemics and, on the whole, good harvests, but point is given to Grenville's remark when we find from price data that the harvest in 1573 had been a bad one, probably due to an unusually wet summer. This led to the average national price of wheat being over

28 per cent above the norm, though prices in the Exeter market were not as high as those over the country as a whole. Grenville's remark implying a sufficiency of people was a daring one for it was well within living memory that the great fear in high places had been that, compared with the traditional enemy, France, England suffered from a shortage, what some called a 'penury', of people, especially those of military age. True, from about the 1520s the great late-medieval drop in population had been halted and slowly but steadily reversed. According to the best evidence we are ever likely to have an overall population of some 2.77 million in 1541 had crept up to 3 million by 1551 and to 3.16 by 1556, the year 1554 (of Raleigh's birth) seeing the annual growth rate for the first time topping 1 per cent. But by 1559, largely as a result of the 'flu, it had fallen to 2.99 million, thereafter to recover, by about 1567, to its level of a decade earlier. No-one could have been confident that the trend thereafter would be consistently upward. The population of Devon had probably increased at a rate higher than that of the country as a whole and indeed in the 1540s the southern fringes, between the Exe and Tamar estuaries, had a settled population of some 20 households to the square mile compared with a national average of little over 10.[3]

It is to Christopher Carleill, writing in 1583, that must go the credit of noting that,

> by the long peace, happy health and blessed plentifulness wherewith God hath endowed this realm...the people is so mightily increased...[4]

The total was probably 3.6 million in 1581 and 3.8 million in 1586 and Richard Hakluyt junior in a famous passage much quoted by demographers, but in fact partly filched from Carleill, wrote in 1584 that, 'through our long peace and seldom sickness...we are grown more populous than ever heretofore.'[5] In fact he was wrong: three and a half centuries before, that is before the Black Death, England had sustained, with only minor crises of subsistence, a population now thought to have been over five million.[6] Moreover by the 1580s many English farmers had the know-how, even if few of them yet acted upon it, to make their land much more productive of both crops and stock. The farmers of Devon, particularly in their use of manure, and largely unconstrained by the communal arrangement of open fields, were among the most efficient in the country.

To the colonial lobby it was manifest that a rising population must place heavy pressure on the country's landed resources. In 1579 John Oxenham told his captors in far away Lima that Grenville and his friends were planning settlements overseas 'because in England there are many inhabitants and but little land'[7] and in 1583 John Hawkins, not perhaps in the first flight of Elizabethan poets, made his contribution as follows:

> So England that is pestered now and choaked through want of ground,
> Shall find a soil where room enough and perfect doth abound...
> But Rome nor Athens nor the rest were never pestered so,
> As England where no room remains her dwellers to bestow,
> But shuffled in such pinching bonds that very breath doth lack,
> And for the want of place they crawl on one another's back.[8]

Yes, indeed, that is very much how it must have looked to a man who, though a Plymothian by birth, was now very much a Londoner, observing daily in the streets of the metropolis the horde of migrants from the provinces, enough to enable her to shrug off the effects of plague and rise from a population of some 60,000 in the early sixteenth century to over 200,000 by 1600, having almost doubled since 1570. Elizabethan London was certainly bursting with humanity.

What however of the countryside, the real world where nine out of ten Elizabethan people still lived? Here almost any parish register will show, both before and increasingly after 1557–9, an annual surplus of baptisms over burials. But we must not be mesmerised by the computers at Cambridge into forgetting that young men and women, those most sought by Raleigh in the mid 1580s, do not spring fully fledged from their mothers' wombs. Philip Amadas who commanded one of Raleigh's reconnaissance ships in 1584 was then a mere 20-year old, that is born in 1564.[9] How many more were there like him? Registrations show that whereas during the influenza years of 1557–9 the number of births in England had been around 90,000 a year, by 1565, probably as the result of earlier marriage following the premature deaths of fathers and elder brothers, the total was over 121,000. It was one of those population bulges of which today universities have to be so aware. Marriages (on the whole) came first and they, we are told, totalled over 40,000 a year in 1560 and 1561 but had dropped back to a more normal 29,000 by 1565. Even allowing for infant deaths there were, indeed, a great many more young adults in England in the 1580s than had been the case for several centuries, lending credence to the younger Hakluyt's reference in 1584 to a realm 'swarming at this day with valiant youths rusting and hurtful for lack of employment'.[10]

To come a bit nearer home let us take a look at the parish register of East Budleigh. This is extant only from 1555, too late to record Walter Raleigh's baptism, but it shows that from a low level of only 5 births in 1558 (in a community of some 250 persons or about 55 families) the total rose over fourfold to 22 and 25 in the years 1565 and 1566. (Incidentally I have news for you: among the mere dozen baptisms in 1559 occurs on 10 May that of an Arthur Barlowe, his father Alexander having married on 26 February, so Arthur, Raleigh's man, if it was indeed he, was a mature chap of 26 when he took ship for Roanoke.[11])

But there was more to lack of employment than a simple increase in population. Studies in depth from various parts of the country, especially in the South East, are revealing not only pressure on the land reflected in rising rents, and more particularly in entry fines for tenancies, but also an increasing polarisation in the size of farms. This took the form of a slow but steady increase in the number of large farms, i.e. those of 50–100 acres, a decrease in the number of farms of 15–30 acres of arable land, the minimum for subsistence, and an increase in the number of smallholdings of 10 acres or less. To put it another way more and more of the available land was falling into the hands of the well-to-do or yeomen farmers. They could afford to provide for an increasing number of sons and daughters by splitting up their land: the small husbandmen could not, or had already done so to the point where there was no sort of a living for any member of the family.[12] John Hooker of Exeter wrote in 1600 of the yeomen farmers of Devon who, once their fines were paid, lived as merrily as their landlords, but he also told of the generality of Devon farms where the farmer and his family eked out their living by sitting by the 'hall-door' carding or spinning wool.[13] Weaving too, though requiring considerably more skill, as well as gear, was very largely a rural, and secondary, occupation, not only in Devon. Other Devon farmers went off in the summer tin-mining and others again went fishing. To the extent that there were these additional secondary occupations available, the farming population of Devon was protected from the chilliest winds that blew across the landscape in other parts of the country, especially in the regions of open fields. In all this the 1580s was probably the crucial decade, the point of no return, and the situation got worse in the 1590s.

By and large landlord pressure was not a serious factor and of course manorial customs still provided some protection, unless tenants could be persuaded to convert their tenures from copyhold (usually 'of inheritance') to leases for fixed terms. Things were most likely to change with the advent of a new landlord. Preliminary findings on the manor of Axminster in east Devon bear this out. The estate was bought by the Duke of Norfolk in 1560. Up to 1560 fines were remarkably stable at 5–6 years' rent but thereafter they rose suddenly to 25 and in the 1570s to 60 times the rent. Perhaps partly as a result, between 1561 and 1585 the number of holdings on the manor under 10 acres increased from 40 to 50 and the total number of tenants from 77 to 97. An increase in population was probably an important contributory factor. Equally significant is the fact that between 1561 and 1595 only about half the tenancies in Axminster remained with the same families, mostly the larger holdings. Most of the copyholds were for years determinable on lives and so the surveyors noted tenants' ages. In 1581 the largest number (40) fell into the 21–30 age group, that is had been born between 1551 and 1560. Only twenty-four had been born between 1561 and 1570. Fourteen were elderly, that is aged between 71 and 90.[14] Where, one wonders, had those who would have been nearing 20 in 1585 gone? As the parish registers are missing we do not know how many of them there were. We know that there was still much 'waste', that is unoccupied, land in Devon, land suitable for what the manorial surveys call 'new takes'. So maybe they had moved to fresh fields and pastures new in nearby parishes. Only further research will tell. Many will have found employment, temporarily or permanently, as farm servants in yeomen's households. For anyone with a mere foothold near the moors there was still good money to be made from sheep-farming. By my reckoning by 1600 there were about three times as many sheep in Elizabethan Devon as people, much as in New Zealand today.

The younger Hakluyt, after noting in 1584 that population was increasing, rather welcomed than deplored the fact. For every thousand men of military age, argued this perceptive clergyman, the country could use five thousand. He went on to make a significant remark which the demographers seem to have missed (though their computers have led them to similar conclusions);

> For when people know how to live and how to maintain and feed their wives and children they will not abstain from marriage as now they do.[15]

Full employment, he seems to be saying, was not only a social necessity but also a political one, and not merely, as the politicians were wont to believe, because hunger breeds disorder.

How could colonisation help the employment situation? It was natural that Grenville, the westcountryman, should have sought to impress the Queen by stating that people in distant lands 'cannot but like well of the use of cloth wherein we most abound'. The Hakluyts, somewhat more realistically, recognised that English woollen cloth no longer enjoyed its traditional pre-eminence abroad, but they thought that, North America being on the whole a cold country, the inhabitants would be delighted with 'any cap or garment made of coarse woollen cloth' and the younger Hakluyt made the very practical suggestion that the colonists should be encouraged to disembark wearing what he called 'Toledo' knitted caps. Having heard of the vast resources of timber on the North American mainland, he pointed to the great expense of importing cyprus chests from the Levant and the Azores, it being the case (well born out in Elizabeth probate inventories) that 'in this age every man desireth to fill his house with all manner of good furniture'. Carpenters and joiners abound, he declared, and one expert could oversee a thousand common millwrights, but whether he envisaged such employment at home or abroad is not at all clear.[16] Hakluyt senior had suggested that men could be employed making wooden spades, 'like those of Devonshire'.[17] Were these those from which the poor tinners were said to drink? What neither considered was how such industry could be financed. The weakness of most new projects in England was lack of capital, as well of course as the attraction they had for monopolists, and, in most cases, their failure to compete in either price or quality with foreign imports. Even Raleigh, with his maritime interest, was later to despair of the statutory enforcement of the growing of hemp for cordage. It could be obtained more cheaply abroad he told Parliament in 1597.[18] If he had been successful in his colonial project the country might well have obtained some cheap raw materials and the diversification of English industry might have been put forward by two hundred years. In fact, of course, when the American colonies did get going early in the next century they discovered, almost overnight it seems, the European market for that very anti-social crop, tobacco.

As long ago as 1549 Sir Thomas Smith in his *Discourse of the Commonweal* had pleaded for the introduction and encouragement of skilled crafts, especially those such as the making of swords, daggers, knives and all tools of iron and steel, pins, thread and various kinds of paper, all of which England obtained very largely from abroad, and at great expense.[19] He would still have been plugging this point in the 1580s for even then only very limited progress had been made in diversifying English industry. Glass-making, under the leadership of an alien, Jean Carré, had been a success story, although the supply of window glass in particular could not satisfy the home market. Almost the only manufactured articles other than woollen cloths which we could offer foreigners were brass cannon, and Elizabeth rather felt that her army could use all that could be produced. More to the point both of these, like clothmaking, were rural industries and did nothing to help occupy the underemployed people in the towns.[20]

What was needed above all was more unskilled or semi-skilled employment. Humphrey Gilbert, writing to his eldest brother John in 1566 in connection with his schemes for the discovery of a North West passage, spoke of the need to

> set poor men's children to learn handicrafts and thereby to make trifles and suchlike which the Indians...do much esteem, by reason whereof there should be none occasion to have our country cumbered with loiterers, vagabonds and suchlike idle persons.

The Peckhams, father and son, said much the same in 1583, referring in particular to the employment of children of 12 or 14 and contending that

> our idle women, which the realm may well spare, shall also be employed on plucking, drying and sorting of feathers...(etc.)

and they seem to have been thinking of employment overseas for they go on to speak of men being engaged in dredging for pearls and hunting whales for oil, as well as working mines and farming, making casks, fishing and lumbering.[21]

So the unemployed were to be shipped across the Atlantic. The younger Hakluyt could refer scathingly to 'the fry of the wandering beggars of England' whom, along with bankrupt merchants, debtors and petty criminals languishing in prison, he suggested might be 'unladen' in the colonies.[22] Here he was mainly echoing official fears of disorder. In fact, as more and more effective use is being made by historians of available records, especially those of the towns, it is being discovered that the great majority of those who moved across England and landed up before the magistrates were not rogues and professional vagabonds, the colourful characters of contemporary literature, but young men in genuine search of employment, or what is called 'betterment', fleeing not so much starvation (though that too came in the 1590s) but lack of land or skilled employment.[23] Carleill made an interesting observation when he wrote in 1583 of

> the great number being brought up during their youth in their parents' houses without any instruction how to get their livings.[24]

How often, in the past, foreign visitors had commented on the peculiar propensity of English families to send their teenage children into other households as apprentices or farm servants. As such they would not earn much but they would at least live reasonably well and learn a skill. The masterless man and the unskilled labourer will always be the first victims of economic crises. How many of these, one wonders, were among the Roanoke colonists?

Identification of Raleigh's recruits, a task still very far from complete, has suggested that most came from London and the Home Counties. This certainly makes sense. All the migration was from country to town and from the highland to the lowland zone, that is towards London and the South East. Poverty was universal but chronic unemployment was largely an urban problem: the landless countrymen just moved into the towns where, even allowing for the sheer weight of numbers, there was more organised charity available. Not until the shortlived commercial crisis of 1586 and the more prolonged one of the 1590s was there widespread rural unemployment, and then largely in those country towns where the population had given up farming and devoted itself almost entirely to clothmaking. Perhaps Raleigh's second contingent of colonists was, on that account, of better calibre than the first. But unemployed weavers were unlikely to be self-reliant colonists.

The promoters of colonies waxed enthusiastic about the benefits to be gained from employment in navigation. Conscripting petty criminals and idlers for the galleys was, of course, an old ploy but what Elizabethan England needed were not brawny oarsmen but skilled mariners, men able to sail a ship beyond the sight of land. The South West had far to go but while John Leland, traversing the peninsula in Henry VIII's reign, had found little but fishermen, by 1560 it was reported to the Privy Council that there were 1270 experienced mariners in Devon alone, by 1570 the number was 1675 and by 1582 over two thousand.[25] Some indication of the move from farming to the sea may be found in the fact that in 1570 the largest number of mariners (123) was said to be in the parish of Stokenham which, though lying in Start Bay, had no haven for shipping. Even neighbouring Blackawton, totally landlocked, had 55.[26]

Drake's circumnavigation of the world in 1577–80 showed how fast Englishmen were learning to take to the deep seas, but probably more to the point was the encouragement given much earlier, during the war with France in the 1540s, to engage in legalised piracy. Walter Raleigh senior was in it up to the hilt, operating from Exmouth.[27] The other gentlemen of Devon could hardly wait for the day when they could prey on Spanish, or indeed any, shipping without any risk of the Queen's displeasure, and many years later Sir Walter Raleigh was to write of the difficulty of persuading other ranks to go to sea unless they could look forward to prizes.[28] We can well imagine that the pressure from those on the lower decks to go after prizes was at least as great as the enthusiasm of the gentlemen. Given a bit of luck the harvest of the sea was gathered in more easily than that of the land. In 1594–5 over 8000 hundredweight of prize sugar paid customs at Plymouth before being shipped to London, an easier way to obtain this precious commodity than working colonial plantations. Plymouth men would have had a dusty answer to the younger Hakluyt's remark that ever since their colonial territories had provided so much employment the people of Spain and Portugal had no longer needed to resort to piracy.[29]

Concurrently, however, the greatest long-term fillip to all that is comprehended in the word 'navigation', that is ship-building, seamanship and the employment of seamen and mariners, at least as far as the South West was concerned, was the Cod Fishery of Newfoundland. Even so the first really large 'catch' to be brought into Plymouth was the 60,000 quintals seized from some Spaniards in 1585 by Raleigh's elder brother Carew and his step-cousin Sir Bernard Drake. By the 1580s several dozen Dartmouth ships were sailing each summer to the Newfoundland coast. Work on the New Quay there, a major feat of reclamation, began in 1584. The following year even the explorer John Davis, a local man, got no support from the leading men of Dartmouth for his search for a North West passage. Fish were the spice of their life. Plymouth too began the construction of her new fish quay in 1572 and in two weeks in September 1595 no less than fifty ships returned there from Newfoundland. Raleigh was never on really good terms with Exeter or Plymouth, but by 1593 even he could describe cod fishing as 'the stay of the West Country' and by then it was probably employing more shipping than either the cloth trade or privateering. It was, moreover, the annual fishing trips which more than anything else prepared westcountrymen for the New England ventures of the next century.[30]

England's need for foreign trade, though it still raises many contentious issues, must needs be dealt with very briefly. By the third quarter of the century the weaknesses were clear. There was the over-dependence on cloth as the one commodity we had to offer and the excessive drain on our spending-power of imported luxuries. Even worse was the disproportionate share enjoyed by the merchants of London and their traditional dependence, both for their export and import trade, on the port of Antwerp. Early in Elizabeth's reign that port was closed to us, and while it is true that alternative cloth marts were found in such places as Emden and Hamburg, we were cut off from supplies of such essential and profitable commodities as far eastern spices, Italian textiles, dye-stuffs, and the wines and dried fruits of Iberia and the Mediterranean.[31] No wonder English eyes were dazzled by the prospects of alternative sources of supply offered by those who had explored the land they called Virginia. Grenville, referring to our hazardous dependence on untrustworthy neighbours (Antwerp was actually sacked by the Spaniards in 1576), seems almost to have been advocating turning our backs on the continent of Europe. Ten years later the younger Hakluyt compiled an horrific catalogue of our hazardous situation, including even the death of the friendly Czar of Russia. He, like Grenville ten years earlier, waxed lyrical over North American supplies, including honey, wax, venison, wildfowl and even silk worms, all of them available good and cheap, mainly for the mere cost of shipment![32]

It is impossible to say what would have been the commercial effects of American colonies in Elizabeth's reign. What we can say is that, having been shown the way in the 1530s by the Hawkins family of Plymouth, English merchants, even some Londoners, had begun to venture further afield and by the 1580s they had already floated many trading associations operative from the Baltic to the Levant and into Africa. Their acquisition of trading facilities rather than the risky establishment of settlements was probably the better bet for the time being. By 1589 even Hakluyt could justly boast of England's world-wide trading contacts. What is more that trade, far more than it had been forty years earlier, was now largely in English hands. Only the merchants of Exeter, smugly protected by their commercial charter of 1561 and their new ship canal, clung to their old outlets across the Channel. That both their export of cloth and their import of wine roughly doubled during Elizabeth's reign was at least some justification for their studied lack of interest in Raleigh's ventures. Even their city chamberlain, John Hooker's, remark in his 'Annals' for 1586, on the occasion of Drake's return with the first colonists, that this had led to many men planning maritime exploits, 'whereby many were undone and themselves in the end never the better',[33] was, at least in the immediate context, only too true.

It is, however, odd that the 'godly' Hooker, one who in his youth had studied under the religious reformers in Strasbourg and had even for a short while tried to tame what he regarded as the totally uncivilised Irish, should apparently have lacked Protestant missionary zeal. There was, in fact, in the literature of colonisation an interesting conjunction of ideas between trading posts in the New World and freedom from religious persecution, at any rate from popery. Until 1569 John Hawkins had tried to do business with

Spain and indeed even after St Juan de Ulua there were English merchants and their factors resident in Spain, but by the 1580s they were finding life under the Inquisition quite insupportable. Both Carleill and Richard Hakluyt senior envisaged colonies as religious havens for merchants.[34] It was quite another thing for good English Protestants to support the idea in 1582 that English Catholic families should be encouraged to cross the Atlantic, the only condition being that at least one in ten of their party should be persons without employment in England.[35] But these, of course, were kith and kin, not the hated foreigners.

But to return to the mission field, each of our writers made much of the point, in the younger Hakluyt's words, that

> western discovery [would] be greatly for the enlargement of the gospel of Christ wherunto the Princes of the reformed religion are chiefly bound, amongst whom her Majesty is principal.

He went on to envisage

> the gaining of the souls of millions of those wretched people, the reducing of them from darkness to light, from falsehood to truth, from dumb idols to the living God, from the deep pit of hell to the highest heavens.[36]

He even saw English colonies as religious refuges for other nationals, no doubt with England's burgeoning alien population in mind.[37]

But the truth is that Elizabethan England was hardly yet in a position to convert the pagan world to Christianity, least of all to Protestantism. Raleigh's colonists included no clergy and if they built a church this is nowhere on record. That in the event they lacked spiritual advisors (who appear very low down even among Hakluyt's list of essential specialists, though he thought they might be useful in maintaining order[38]) should not however surprise us, for Elizabethan England was woefully short of clergy and especially of what were called 'preachers', men able to expound the faith, let alone keep up intellectually with a Harriot or a White. In 1586 it was reported that only one fifth of English parishes possessed a priest capable of instructing his flock and even in 1603 the number was barely one half. (In Devon the ratio was about 1 in 6 in the mid 1580s and 1 in 3 by the end of the century, and this in spite of a succession of bishops who were rather partial to preachers provided they did not stir up trouble.) Moreover most congregations were content that it should be so, believing as most English people still did in salvation by good works rather than in the more intellectually-demanding doctrine of justification by faith.

It is indeed being increasingly recognised that in general the ordinary people of England were slow to accept the Reformation, preferring, like their Queen, to hold on as far as they could to traditional beliefs and practices.[39] Puritans, 'the hotter kind of Protestants', with their as yet somewhat un-English liking for reading the Bible and listening to long sermons, were a small minority, mere pockets in a predominantly 'C. of E.' environment. Sectarianism hardly existed except the small circle around Robert Browne in East Anglia, and his departure to Middleburg was in no sense the beginning of a flight from an oppressive religious régime. In the South West the Puritans were confined to certain parishes dominated by 'godly' gentry, notably those like the Chichesters of Raleigh near Barnstaple who had gone into exile under Mary. Having gone abroad very largely to escape Spanish influence at Court, many had come back converted, but even in Mary's reign there were many gentlemen who hated the Spaniards more than their religion. Had the worst happened and the Spaniards toppled the Elizabethan régime it would have been largely the gentry who, once again, would have spear-headed the flight abroad, this time perhaps to the New World. They would have gone, with as many hangers-on as they could muster, as political rather than religious refugees. The majority of ordinary people would no doubt once again have accommodated themselves to official directions, and been thankful that they still had some of their traditional church ornaments hidden in haylofts. Even for Catholic recusants religious freedom was no good reason for leaving Elizabethan England.

But the interest of the gentry in colonies went far deeper than either religion or politics. On the whole the colonial literature did not stress the advantages of extending the Queen's dominion, the authors knowing full well that the Queen had no territorial ambitions. She had after all preferred the more strictly maritime project of Francis Drake to that of Grenville. One of John Oxenham's companions had told his captors in 1579 that the Queen was the great obstacle to

English colonial endeavours but that if she should die the floodgates would be open.[40] Even in 1584, when England was virtually at war with Spain, Raleigh's patent confined him strictly to territory not yet occupied by any Christian prince. Otherwise it merely provided that the colonists would remain English denizens and that such laws as might be devised would as far as possible conform with the English common law. Was the Queen conned by her favourite courtier? As proprietor of what would have been virtually the whole of the present United States of America, Raleigh, like Gilbert before him, was to hold by homage only, and merely to pay a royalty to the Queen of one fifth of any precious metals. He was to enjoy absolutely free disposal of his land, with not so much as a mention of the Queen's rights of prerogative wardship over minors and her disposal of the marriage of daughters which was such as aggravation to landowners in England.[41] Did the Queen, or the Lord Chancellor (she did not have a proper one until Hatton took over in 1587), know what she was granting? Or was it a calculated gamble, costing her nothing and ensuring Raleigh's continued service at home? It was tantamount to a revival of the old English (and Irish) palatinates which astute early Tudor monarchs had striven so hard to extinguish. We can obtain some idea of how Raleigh might have proceeded to exercise his North American lordship from the provisions made by Gilbert in 1582 for the benefit of his widow and children. His trustees were to establish an almost feudal régime of 'seignories', whose grantees were to be obliged to provide one soldier fully equipped for war for forty days in the year for every five thousand acres. Lessees for lives were to enjoy hedgebote and ploughbote and to pay heriots.[42] None of this will surprise anyone who has looked at Sir Thomas Smith's scheme in the early 1570s for the plantation of Ireland, his clients being, of course, very largely the Grenville west country cousinage. But Smith and his friends were to hold their estates by knight service as of the (royal) earldom of Ulster.[43]

What a prospect in 1584 for Walter Raleigh! If his father and elder brothers had made any real money out of privateering they had not invested the proceeds in land: rather had they sold parts of their modest inheritance to buy ships. As the youngest of four sons Walter had little hope of more than a very modest patrimony. In 1561 he was given a residual interest in the manor of Colaton Raleigh in east

Devon, an estate which he eventually possessed, but only in 1597 on the death of his eldest brother George. When in the 1580s he designated himself 'of Colaton Raleigh', this can only have been in respect of a small estate called Hawkerland in that parish which he had somehow acquired, directly or indirectly, from Gertrude Courtenay, dowager marchioness of Exeter.[44] At this time he owned no other land in England.

Raleigh was knighted in January 1585, being then already member of parliament for Devon, both of these unusual achievements for a virtually landless gentleman. Later that year he was to succeed the deceased Francis Russell, second earl of Bedford, as Lord Warden of the Stannaries, High Steward of the Duchy of Cornwall and Lord Lieutenant of Cornwall. As such he would enjoy power and patronage, but no landed inheritance, without which there was no future for his line. Many still alive remembered that in 1539, John Russell, the later first earl of Bedford, on becoming Henry VIII's chief man in the South West, filling the power vacuum left by the execution of Henry Courtenay, marquis of Exeter, had received from the crown a very large estate in Devon.[45] Not so Raleigh in 1585, this not being Elizabeth's style. Even had he invested what cash he had in English land rather than in colonial ventures, land suitable for gentlemen, that is manors and other revenue-producing property, was no longer readily available, even for purchase. As is well known Raleigh had tried in vain in July 1584 to buy from Mr Richard Duke the farm of Hayes in East Budleigh where he had been born, but even had he succeeded he would have been tightly hemmed in by the Dukes and his cousins the Drakes, families which between them had long since bought up virtually the whole of the extensive former monastic estates in that corner of Devon.[46] Only the bulging pockets of a Francis Drake could have dispossessed Grenville of Buckland Abbey but I suspect that Raleigh had no wish to be seated so far west. His best hope, of course, was to marry an heiress, but so far he seems not to have been the marrying kind, and indeed it is difficult to think of a suitable candidate, even among his vast network of landed cousins.

What I am suggesting is that for ambitious young gentlemen, especially younger sons, colonial territory offered opportunities not only of immediate wealth but also of the establishment of a landed dynasty not easily available in England. It was an aspect of colonisation to which the promoters, except for the Catholic Peckhams, devoted very little attention, perhaps for fear of alarming the Queen who, at least until Raleigh's faux-pas of 1592, wanted him by her side. Raleigh's somewhat equivocal commitment to his American colonial schemes after 1585 still requires a convincing explanation. Was it perhaps that Ireland increasingly seemed to offer better chances of landed wealth? By 1589 he had settled more colonists in Munster than ever he even tried to settle in America.

NOTES

1. *New American World: a documentary history of North America to 1612*, ed. D. B. Quinn, 5 vols 1979. Vol. III contains all the relevant texts from 1576, with authoritative introductions.
2. Public Record Office, State Papers Domestic (SP12), vol. 95, nos 63–4, printed in R. Pearse Chope, 'New Light on Sir Richard Grenville', *Devonshire Association Transactions*, xlix, 1917, pp. 216–22, 237–41.
3. E. A. Wrigley and R. S. Schofield, *Population History of England 1541–1871*, 1981, p. 531; P. Slack, 'Mortality Crises and Epidemic Disease in England 1485–1610', in C. Webster(ed.), *Health, Medicine and Mortality in the Sixteenth Century*, Cambridge, 1979, pp. 40–41; P. Bowden, price tables printed as appendices to J. Thirsk (ed.), *Agrarian History of England 1500–1640*, Cambridge 1967, p. 819; W. G. Hoskins, 'Harvest Fluctuations and English Economic History, 1480–1619', *Agricultural History Review*, XII, 1964, p. 46 and J. Sheail, 'The Distribution of Taxable Population and Wealth in England during the early sixteenth century', *Transactions of the Institute of British Geographers*, 1972, p. 119.
4. *New American World* III, p. 31.
5. *ibid.*, p. 82.
6. J. Hatcher, *Plague, Population and the English Economy 1348–1530*, 1977, p. 68.
7. R. Pearse Chope, *op. cit.*, p. 245.
8. *New American World* III, pp. 36–7.
9. J. L. Vivian, *Visitation of the County of Devon*, Exeter 1895, p. 12.
10. Wrigley and Schofield, *op. cit.*, pp. 537, 553 and *New American World* III, p. 119.
11. Devon Record Office, East Budleigh Parish Register, 1555–   .
12. See, for example, M. Spufford, *Contrasting Communities: English Villagers in the sixteenth and seventeenth centuries*, Cambridge 1974, pp. 46–57.
13. 'Hooker's Synopsis Chorographical of Devonshire', ed. W. J. Blake, *Devonshire Assoc. Trans.*, xlvii, 1915, pp. 341, 346. A more complete edition of this important text is in preparation.

14. From an unpublished paper by Mrs Angela Langridge and quoted with her kind permission.
15. *New American World* III, p. 84.
16. *ibid.*, pp. 122, 103.
17. *ibid.*, p. 68.
18. From the parliamentary diary of Sir Simon D'Ewes, quoted in P. W. Hasler, *The House of Commons 1558–1603*, 1981, II, p. 275.
19. M. Dewar (ed.), *A Discourse of the Commonweal of this Realm of England*, 1969, p. 122.
20. Joyce Youings, *Sixteenth-Century England*, 1984, pp. 238–44 and bibliography on pp. 406–7.
21. *New American World* III, pp. 21, 50.
22. *ibid.*, pp. 120, 123.
23. See in particular A. L. Beier, 'Vagrants and the Social Order in Elizabethan England', *Past and Present*, 1974.
24. *New American World* III, p. 31.
25. M. Oppenheim, *The Maritime History of Devon*, Exeter 1968, pp. 38, 40.
26. PRO, State Papers Domestic (SP12), vol. 71, no. 75.
27. M. J. G. Stanford, 'The Raleghs take to the Sea', *Mariner's Mirror*, 1962.
28. In 1585 Raleigh was licensed to impress men in Devon, Cornwall and Bristol, but he seems to have had to make up his complement with alien seamen: D. B. Quinn (ed.), *The Roanoke Voyages*, Hakluyt Society, 1955, pp. 144–5, 151, 834–5.
29. T. S. Willan, *Studies in Elizabethan Foreign Trade*, Manchester 1959, p. 81 and *New American World* III, p. 82.
30. P. Russell, *Dartmouth*, 1950, chap. III; H. A. Innis, *The Cod Fisheries*, Toronto, 1954, *passim* and for a recent summary, N. C. Oswald, 'Devon and the Cod Fishery of Newfoundland', *Devonshire Assoc. Trans.*, 115, 1983.
31. R. Davis, *English Overseas Trade 1500–1700*, 1973 and D. M. Palliser, *The Age of Elizabeth*, 1983, chap. 9, pp. 278–91.
32. *New American World* III, pp. 74–6. See also Carleill, *NAW* III, pp. 28–9.
33. Davis, *op. cit.*, *passim* and Willan, *op. cit.*, p. 80. For Hooker's comment see D. B. Quinn, *The Roanoke Voyages*, Hakluyt Soc. 1952, p. 313, but Hooker takes a more favourable view in his dedication of a book to Raleigh later the same year: *New American World* III, pp. 312–3.
34. *New American World* III, pp. 29, 63.
35. D. B. Quinn, *England and the Discovery of America*, 1973, p. 373.
36. *New American World* III, pp. 71, 73. See also what the Peckhams wrote (*ibid.*, p. 53) which Hakluyt roughly copied.
37. *ibid.*, p. 120.
38. *ibid.*, p. 123. Cf the plans for Ireland in the 1570s where church building had a high priority and lands were to be set aside for the maintenance of clergy, parish clerks and school masters: see note 43 below.

39. For current thinking see the chapters by P. Collinson and C. Haigh in C. Haigh (ed.), *The Reign of Elizabeth I*, 1985 and P. Collinson, *The Religion of Protestants*, Oxford 1982. For Devon see I. Gowers, 'Puritanism in Devon', unpublished M.A. thesis, University of Exeter 1970.
40. R. Pearse Chope, *op. cit.*, p. 246.
41. *New American World* III, pp. 267–70.
42. D. B. Quinn (ed.), *The Voyages and Colonizing Enterprizes of Sir Humphrey Gilbert*, Hakluyt Soc., 1967, pp. 266–78.
43. M. Dewar, *Sir Thomas Smith: a Tudor Intellectual in Office*, 1964, pp. 156–70.
44. Details will be found in my booklet on Sir Walter Raleigh and the South West of England to be published shortly by the North Carolina Department of Cultural Resources.
45. Joyce Youings, 'The Council of the West', *Trans. Royal Historical Soc.*, 1960 and D. Willen, *John Russell, first earl of Bedford*, 1981.
46. Joyce Youings, *Devon Monastic Lands: Particulars for Grants 1536–58*, Devon and Cornwall Record Society, New Series I, 1954, *passim*.

4

The Lost Colonists

David Beers Quinn

The problem of the Lost Colonists begins with the first colony brought to Roanoke Island in 1585 and remaining there until brought back to England by Sir Francis Drake in July 1586. What should Sir Walter Ralegh do to follow up their return? First, he must wait to see what the supply ship he had sent too late to help them would report when it returned: second, he must also await Sir Richard Grenville who had gone out with additional colonists but, again, too late to help the first colony. During the months between July and Grenville's return in December, reporting that he had left only fifteen men to hold Roanoke Island in the interim, much hard thinking had taken place. Ralph Lane, the governor in 1585, was of the opinion that Roanoke Island could not act as a privateering base: in any case the Indians were now hostile. A new base with a deep harbour might be found in Chesapeake Bay to the north. Thomas Harriot was able to give a full report on the area and to exhibit a map compiled by himself and John White the artist, showing a deep water entry to Chesapeake Bay. Harriot's report on the natural resources of the area could point to Indian-grown commodities like maize, which were valuable for sustaining a colony, but all his ingenuity could not discover signs of gold (though some of copper) and little hope of growing Mediterranean-type commodities in the latitudes he explored.

White appears to have taken an original and promising initiative

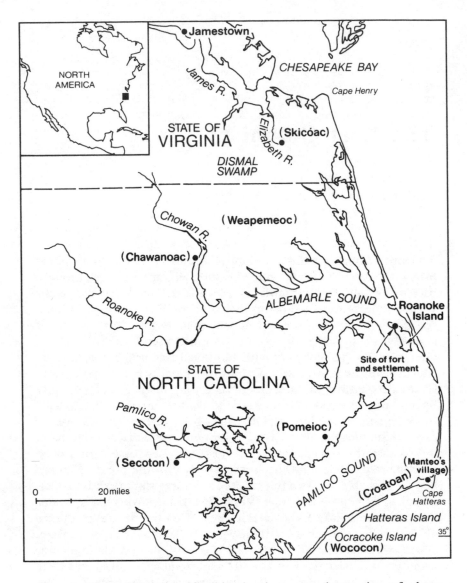

Fig. 4. Map of Ralegh's Virginia showing approximate sites of places mentioned in the text. Inset: map of North America showing location of Ralegh's Virginia.

60

in these discussions. He and Harriot had lived over the winter of 1585–6 with the Chesapeake Indians, inland from Chesapeake Bay at Skicóac on the borders of the Great Dismal swamp. If the Indians of the Roanoke area had largely turned hostile, could not a real colony of men, women and children settle alongside the Chesapeake Indians and begin a new life there, leaving aside the question of a privateering base to the decision of Ralegh and Grenville at a later stage? Well before Grenville's return White had put this project under way. Ralegh was sympathetic and may have done much to put the promoters in touch with his contacts in the City of London, but his interests were focussed on a privateering base and this type of settlement colony in America had only a marginal interest for him (though he was engaged in preparing some such sort of colony in Munster). The result was that White (with Harriot, we believe) mustered a fairly wide degree of support in the City. This did not come, we are almost certain, from any of the great merchants, but from members of the minor craft guilds of London (White was a member of the Painter-Stainers') who wished to leave the cramped and limiting conditions of city life for the promise of land, the large estates of five hundred acres offered to them in America. Agents for this promotion appear to have been active also in Essex and perhaps in Hampshire among smallholders greedy for land and for opportunities they could not get in England.

So, by the beginning of 1587 there had emerged the Governor (John White) and Assistants of the City of Ralegh in Virginia, empowered to create a new society in North American land. We know little of their makeup, though White's son-in-law, Ananias Dare, was a member of the Tilers' Company. But they also included Simon Fernandes, a man of dubious character, the Portuguese pilot who had led the reconnaisance expedition of 1584 to Roanoke Island and who pretended, we believe, to know the Chesapeake as well. Ralegh gave this body status by arranging for a grant of arms for the city, governor and twelve assistants to be made on 7 January 1587. White may have set his sights on as many as two hundred and fifty persons but then lowered them to one hundred and fifty, and finally set sail with only some 118 individuals, men, women and children, in April 1587, the ships, we believe, being supplied by Ralegh, along with £100 to help with immediate costs.

When they reached the Carolina Outer Banks, behind which Roanoke Island lay, on 22 July, intending to make contact with Grenville's holding party on the island, they were faced with an ultimatum. Fernandes would not convey them to Chesapeake Bay: they must be landed on Roanoke Island. The alleged reason was that he intended the ships to sail off privateering and it was too late to sail further along the coast, but this was a lie since they remained for a month at the Outer Banks. The real reason is unknown. White acquiesced in the ultimatum, which was backed by the seamen, for the sake of his helpless passengers. They settled temporarily into the quarters vacated by the first colony, but found no trace of Grenville's men (who had been chased away after an Indian attack and had sailed to almost certain shipwreck many months before). White was faced with a crisis. The settlers were prepared to go to the Chesapeake (they had a pinnace which could be used for this purpose) and this is shown by the handing over of Roanoke Island and its environs to Manteo, the Indian leader who had been twice in England and who was baptized as a Christian at the same time as Virginia Dare, the first child to be born in English America and White's grandchild. But what of the future supplies for the colony and those other colonists who had arranged to come out later? White must go back for supplies and reinforcements. He could not be certain where he would find the main body of colonists when he came back since they were going on trust into the wide Chesapeake area. A party must await his return. Reluctantly at length he set sail in one ship which intended to sail home while the other was seeking prizes. He was never to see the colonists again: they became the Lost Colonists of history and legend.

At the end of 1586 White was faced in England with tremendous difficulties. He found Ralegh and Grenville preoccupied with both English and Irish responsibilities. He did not manage to get to sea as soon as he had promised and hoped to be able to do. On the other hand, he found Grenville preparing an expedition to go to the Chesapeake to establish a quite independent privateering base and had to sit out the winter between London and Bideford in Devon, eventually collecting a few additional settlers and obtaining stores, partly perhaps from London backers; partly, it is not unlikely, as a gift from Ralegh and Grenville. But the Grenville expedition was stopped by royal order as it was about to sail. It must join the fleet at Plymouth which was preparing to encounter the Spanish Armada, now at last ready to descend on English shores. In spite of this hard blow White did slip away in April 1588 with two small pinnaces with some stores and a handful of reinforcements, but he found the sea infested with pirates and his own sea captains only too ready to join in the melée. To their cost it was the *Brave* and the *Roe* which were looted and forced to limp back to England with casualties, White suffering three wounds from which he must have made a fairly quick recovery. The Armada was driven away with heavy losses but neither Ralegh nor Grenville could help White directly. However, Ralegh did get his business manager, William Sanderson, to put together a syndicate which included some of the richer city merchants to underwrite the colony on patriotic grounds. This was in March 1589, but no ships sailed for Roanoke in that year. Either there was not enough money or there were no ships available. Technically there was a ban on ships leaving on overseas voyages but others evaded it or were licensed to do so. The crux was probably that White could not venture again unprotected across the Atlantic and that no privateer would take the risk of carrying noncombatants, especially women and children, through the Caribbean. Finally, in 1590, Ralegh did arrange for White (with some heavy artillery and other supplies in the hold of the leading ship) to sail in the *Hopewell* and to call at Roanoke, while Sanderson and the syndicate equipped a supply vessel, the *Moonlight*, to accompany the privateers.

In the event, Captain Cocke of the *Hopewell* would take White alone on board; the *Moonlight* was not ready when he sailed but a rendezvous in the Caribbean was arranged and, in fact, effected. But before any visit to Roanoke could take place fierce fighting and prizetaking had to be endured. Finally, on 15 August 1590, the *Hopewell* and *Moonlight* anchored off the Outer Banks and the search for the Lost Colonists began. White expected to find a party awaiting him on the island, prepared to lead him to the main colony, now, he expected, installed alongside the Chesapeake tribe some fifty miles to the north. But a storm had just passed, the waters were rough, a boat overturned, losing the master of the *Moonlight* among others, but eventually a search was made of Roanoke Island and revealed only puzzles and no Englishmen. The older buildings were gone but a new palisade had been erected, with no buildings inside, though with cannon and iron bars left behind there. Finally a trench was found to have had buried in it some chests belonging to White and others, but these had been discovered and looted by the Indians. The only ray of hope was the word CROATOAN cut in a post, which meant to White, though he was puzzled by it, that *all* the colonists, not merely the party he hoped had waited for him (for three years!), had gone to Croatoan Island where Manteo's people scratched a bare living from the widest part of the Outer Banks. But continuing bad weather and loss of anchors obliged White to sail off without getting to Crotoan, hoping against hope he could get the ships to winter in the Caribbean. In the end he sailed home in the *Hopewell*, reaching Plymouth on 24 October. By the end of 1590 he was in London, a defeated man, his colonists lost for ever. He failed to find any means of sailing back and, in a sombre, quietist frame of mind, eventually moved to Ireland and took up a tenancy in the Munster plantation, a colonist indeed there, but an unwilling one. And so he disappears from our sight, though his drawings of the Indians and their way of life, made in 1585–6, remain his lasting memorial.

The colonists were almost forgotten in the war years which followed, but not quite. A few speculations about their fate or survival are found in the literature between 1597 and 1603, but nothing of consequence. Ralegh began sending small reconnaissance vessels across the Atlantic, possibly about 1599. Samuel Mace in 1602 did bring back some American medicinal plants and other goods from what we think was the southern part of the Outer Banks below Cape Hatteras, but he did not reach Croatoan. In 1603 Ralegh arranged for Bartholomew Gilbert and Mace to go with two vessels to search Chesapeake Bay. Gilbert failed to find the entrance to the Bay and went ashore north of it, to be killed, with some of his men, by a party of hunting Indians. Of Mace we have no direct evidence, but in September 1603 some North American Indians were showing off their paces in a canoe in the Thames, just below the Strand, and it is almost certainly in that year that some Indians were taken from the York river, farther north in Chesapeake Bay. (By then Ralegh had lost all his American rights which he had sustained thus far on the assumption that the Lost Colony still existed, and was a prisoner in the Tower.) Some news of the Lost Colony may have come with the Indians. The fact that in 1605 *Eastward Hoe* made fun of the Lost

Colonists on the London stage suggests that some ballad or broadsheet had proclaimed their continued survival since otherwise the reference would have been unintelligible to the audience. These conclusions are largely speculative, as its the view that the Virginia Company (the London division of which was intended to colonise Chesapeake Bay) received its charter in April 1606 partly as a result of the revelations (if they are such) of 1603. There is no doubt that the Jamestown settlers from May 1607 onwards had instructions to look for and find the Lost Colonists if they still existed.

It may well be that they very nearly established what had happened to them at their very first landing, a little westward of Cape Henry (which they had just named), if indeed the Lost Colonists had been, as seems almost certain, massacred by Powhatan, the ruler of much of the Virginia Tidewater, shortly before the arrival of Captain Newport and his pioneer colonising party. Newport was attacked from Cape Henry by Indians who he was later told were Chesapeakes and they caught a glimpse of some of them after the initial attack in which some Englishmen were wounded. But a party traversed a wide area to the south and west of Lynnhaven Bay without seeing any Indian settlements (though there are several on the White-Harriot map), finding open country but being eventually stopped by a curtain of smoke and fire which they thought had been created by the Indians who had attacked them. The expedition returned to their ships and went ahead, up the James River, and duly established the first party of settlers at Jamestown. But as soon as this was done Newport took a vessel upstream to explore the upper reaches of the James River. Near the point where its broad course was interrupted by the rapids and rocks of the Fall Line he encountered an Indian gathering. Its leader proclaimed himself to be Powhatan and claimed to be ruler of all the Indian groupings around the Tidewater, excepting only the Chesapeakes, who were enemies. He advised the English not to settle with them and they agreed not to do so, some still bearing scars inflicted on them at the first landing. This whole episode was farcical. The Indian leader was not Powhatan but one of his sons, Rarahunt, and the fulminations against the Chesapeakes appear to have been precisely aimed at preventing the English from finding out that the Lost Colonists had recently been wiped out, along with their hosts, the Chesapeake Indians, a result of Powhatan being told by his priests that these people, the Lost Colonists and their hosts, were about to destroy his regime. This would identify the hostile Indians of the Cape Henry area as Powhatans, recently engaged in this bloody exercise. Something of this version of what happened in the earliest stages of English exploration in 1607 is speculative, since we do not know how much any Englishmen could understand of what Rarahunt said, but it is clear that the massacre had taken place, though precisely when—in 1606 perhaps, or even in April 1607 as the colonising fleet was first sighted—will probably never be known.

One obstacle in finding out what actually happened is that Captain John Smith, in the record of the explorations he made between 1607 and 1609 on behalf of the Jamestown colony, did not record anything about the massacre. On the other hand, he told the Reverend Samuel Purchas, who was writing his 'Virginia's Verger' in 1623: 'Powhatan confessed to Captain Smith that he had been at their slaughter and had divers utensils to show.' Moreover, in 1625 Purchas annotated his great compilation, his *Pilgrimes*, in which Smith had given him much help, with the statement:

> Powhatan confessed that he had been at the murder of that colony and showed to Captain Smith a musket barrel and a bronze mortar and certain pieces of iron which had been theirs.

Smith too had said nothing in his *Map of Virginia* (1612) about his visit to the Elizabeth River, some miles up which Skicóac, the Chespeake chief town, had stood. In 1624 in his *Generall historie* he did say that he ventured into that river in September 1608, sailing up six or seven miles and seeing two or three garden plots with houses, but no Indians, before he turned back. With our present knowledge it might seem that the Indians hid because they did not wish to be identified as new Powhatan settlers and not Chesapeakes. However, Smith had earlier told of sending an expedition southwards in a search for the Lost Colonists: it penetrated into the area where the colonists of 1585–6 had been and got word of four Englishmen who had been seen, though no contact was made with them.

The chief source for our knowledge of what happened is William Strachey's 'Historie of travell into Virginia Britania', largely written in the Jamestown colony in 1611. According to Strachey,

> His Majesty [James I] hath been acquainted that the men, women and children of the first plantation at Roanoak were by practice and commandment of Powhatan (he himself persuaded thereunto by his priests) miserably slaughtered without any offence given him.

Moreover, he tells us that the Chesapeake tribe had been wiped out by Powhatan and, of the Lost Colonists, that for 'twenty and odd years [they] had peaceably lived and intermixed with those savages [the Chesapeakes] and were off his [Powhatan's] territory.' That such information was sent to England at the end of 1608 or at the opening of 1609 is verified by the instructions given to Sir Thomas Gates when he was leaving with a large expedition in May 1609. Several passages leave no doubt that Powhatan was to be punished for his actions against the Lost Colonists. One stated:

> For Powhatan and his *weroances* [tribal leaders] it is clear even to reason, besides our experience, that he loved not our neighbourhood, and therefore you may in no way trust him, but if you find it best not to make him your prisoner, yet you must make him your tributary and all other his *weroances*, first to acknowledge no other lord but King James and so we shall free them all from the tyranny of Powhatan.

This involved destroying the precarious but formidable unity of the Powhatan Confederacy, and removing most of the high chief's authority. A second instruction implies (and Strachey asserts) that the priests were the inspirers of Powhatan's massacre.

> We think it reasonable you first remove from them their *iniocasockes* or priests by a surprise of them all and detaining them prisoners...And in case of necessity or convenience, we pronounce it not cruelty nor breach of charity to deal more sharply with them and proceed even unto death with these murderers of souls and sacrificers of God's images to the Devil.

They were in fact to be killed if it suited the governor of the colony. Strachey repeats the greater part of this official instruction in his book. The order was not, however, to arrive in Virginia until 1610 as the *Sea Venture* carrying the governor, Strachey, and the instructions were held up by the wreck of their ship on Bermuda for nearly a year. In 1610 the colony was too weak to try to implement the instructions and, though they were sent out again in 1611, Powhatan was left unmolested. Keeping the colony alive was the most the English could do.

Strachey had spoken of the massacre as having taken place 'at what time this our colony (under the conduct of Captain Newport) landed within the Chesapeake Bay', namely in April 1607, although it could have taken place somewhat earlier. The 1609 instructions showed that detail about the circumstances had already reached England by May of that year. It was ordered that an expedition should go south to a place called 'Peccarecomicke' and it was said:

> You will find four of the English alive, left by Sir Walter Rawley, which escaped the slaughter of Powhatan of Roanoke upon the first arrival of our colony.

It was further stated that, 'they live under the protection of a weroance called Gepanocon, by whose consent you will never recover them.' This statement shows that some of Smith's attempts to search to the south for the colonists had borne some fruit. Strachey was able to add to it a little, saying that 'at Ritanoe the *weroance* Eyanoco [the Gepanocon above?] preserved seven of the English alive, four men, two boys and one young maid, who escaped and fled up the river of Choanoke [Chowan] to beat his copper, of which he hath certain mines.' We thus have seven survivors living, it appeared, among the Chowan Indians of North Carolina, where three men left behind by Lane in 1586 may also have gone. Further expeditions, the last as late as 1622, could find no trace of the survivors and they pass out of recorded history, but leave these faint traces of North Carolina's beginnings.

The larger party whom White expected to find on Roanoke Island in 1590 was never located. It is not improbable that they had completed their palisade in June 1588 when a Spanish ship on a reconnaissance put in at Roanoke Island and discovered the landing place which led to the settlement, though not the palisade itself. It seems not at all unlikely that they thereupon abandoned the island completely, making perhaps sporadic visits to it in their pinnace (which must have survived the transportation of the main colony to the vicinity of Skicóac in 1587), but lived on Croatoan Island alongside Manteo's people. They would naturally expect White to have read their signs on Roanoke Island when he did come and would look out for him on Croatoan, near modern Cape Hatteras, but as he never appeared we can suggest that they assimilated with the Indians, perhaps, since the resources of the island were so limited, migrating in part at least to the mainland. The Hatteras Indians, seen by John Lawson in 1701, claimed to have European ancestors and certainly had some indications of European features. All else about the Lost Colonists, and there is much, is myth.

It is necessary to suggest some possible explanation of John Smith's failure to deal in his published works with the end of the Lost Colonists. Of course the story of their massacre was bad propaganda for Virginia colonisation which he continued to promote, but there is more to it than this. In 1606 James I had created a general council for Virginia, which was an official body superior to the Virginia Company, set up for primarily commercial and settlement objectives. In 1608 Christopher Newport had recognised Powhatan as a tributary of the English crown and had even crowned him as a subsidiary monarch on behalf of King James. Then, at the end of 1608 or the beginning of 1609, Smith heard from Powhatan himself that he had been responsible for the massacre of the Lost Colonists. What was he to do? The Virginia Company would be so embarrassed by this knowledge that Smith could only communicate it directly to King James himself. It is Strachey who says the letter was sent, but only John Smith as Governor of the colony at the time could have sent it. The action to be taken was something beyond the Company's competence. It was therefore the king, through his royal Council for Virginia, not the Company, which issued the instructions of May 1609, revealing the fact of the massacre and providing for the punishment of Powhatan and his priests. Smith, when he returned to England in 1609, received little credit for his work in the colony, and it is suggested that one reason for it was that he went over the head of the Company by approaching the king directly. It is notable that his famous *Map of Virginia* of 1612 was published in Oxford, away from pressure by the Virginia Company on the Stationers' Company which controlled the licensing of books. But even there it seemed unwise to come clean about the Lost Colony so that the visit to the Elizabeth River was suppressed. After 1622 when a fresh massacre had killed some hundred of the colonists, Smith felt he could be frank at last. His *Generall historie* in 1624 revealed the abortive visit to the Elizabeth River: his private communications to Samuel Purchas, duly published in 1625 (after the Virginia Company had been suppressed), told what Powhatan had boasted of to him so long before. This is a tentative reconstruction, but a likely one, of one of the mysteries of the failure to reveal what happened to the colonists until long after. Strachey was unable to get his book published as it overlapped too much with Smith's *Map* and was, in any case, never completed. It first appeared only in 1849.

Why, apart from the detective story interest which the Lost Colony excites, and the place its story and its myths play in the background of modern Northern Carolina, should the Lost Colony have some importance in the history of English settlement in North America? It was, it may be argued, the first attempt to create in America an autonomous community of English people who would be largely self-supporting and thus antedated the Pilgrims of 1620 by over a generation. In the second place, it was planned (if this interpretation is correct) as a community which was based on close cooperation with the Indians. White and Harriot had spent the winter of 1585–6 with the Chesapeake Indians and concluded that it was with them that a friendly partnership could be built up on the fertile lands, not yet thickly populated, on the fringes of the Great Dismal swamp. The little evidence that we have, from Strachey alone, indicates that they achieved this object. The two communities, that of Skicóac, we think, and the City of Ralegh, coexisted and may before the end have even coalesced. This was to be unique in the early history of English enterprise in North America, and it is a tragedy that we do not know more about it. Finally, the Lost Colony, even if its fate was bitter and final, did, through its survival for twenty years, act as a link between the Roanoke colonies of 1585–87 and that of Virginia from 1607 on, so that there was some degree of continuity in English settlement between the experiments of Elizabeth's reign and the more permanent achievements of her successor's. What can be reconstructed on all these points is clearly tentative and to some degree speculative, but it gives at least a coherent story for the first time of what may well have happened.

SELECT BIBLIOGRAPHY

D. B. Quinn (ed.), *The Roanoke Voyages, 1584–90* (Cambridge, The Hakluyt Society, 1955)

D. B. and A. M. Quinn, *The First Colonists* (Raleigh, N.C., Department of Cultural Resources, 1982)

William Strachey, *The Historie of Travell into Virginia Britania*, ed. L. B. Wright and V. Freund (London, The Hakluyt Society, 1953)

Smith, John, *The Works*, ed. Philip L. Barbour, 3 vols (Chapel Hill, North Carolina, 1986)

S. M. Kingsbury, *Records of the Virginia Company of London*, vol. III (Washington, D.C., Library of Congress, 1935

D. B. Quinn, *The Lost Colonists: their fortunes and probable fate* (Raleigh, Department of Cultural Resources, 1984)

D. B. Quinn, *Set Fair for Roanoke* (Chapel Hill, University of North Carolina Press, 1985)

5

The Americanization of Raleigh

H. G. Jones

Ninety years ago the English historian James Anthony Froude wrote, 'of Raleigh there remains nothing in Virginia [i.e., North Carolina] save the name of the city which is called after him.' Well, colleague Froude, wherever you are, this paper is for you.

Had Froude been in North Carolina when he made that assertion in 1895, he would have substituted the word 'little' for 'nothing' and reported that a surge of interest, beginning with the tercentenary of the Amadas and Barlowe expedition, promised to resurrect the name of Sir Walter Raleigh. Even so, the historian could not have foreseen that the Shepherd of the Ocean would within a century become so thoroughly Americanized that you would bring me across the Atlantic to explain how today Raleigh may be better known by my countrymen than by yours.

Immortality tends to crown success and spurn failure, so the very fact that Raleigh's colonies failed while those at Jamestown and Plymouth Rock succeeded helps explain why for three centuries little was heard of the Roanoke ventures. Moreover, except for a few accounts by Elizabethans, the drawings of John White, and the surviving earthworks on Roanoke Island, we would not be here today, for the story of the expeditions sent to North America by Walter Raleigh would be comparable to the oral traditions of the

Norsemen, with precious little historically reliable evidence to support them. If that had been the case, one cannot help wondering to what cause David Beers and Alison Quinn would have devoted their careers in the twentieth century!

It is not my purpose today to discuss *English* interests in Sir Walter Raleigh, but rather to trace his gradual adoption by Americans over a period of four centuries. Still, we must remember that the overwhelming majority of Americans were themselves English in nationality for half of the period under review, and there are those of us who still proudly refer to our nation as English America.

The land distance between John White's Virginia of 1587 and John Smith's Virginia of 1607 was less than that from Exeter to Land's End, but that distance was deceptive, for hundreds of square miles of marsh between Jamestown and Roanoke Island made land travel difficult, and the treachery of the shifting sands of the outer banks was already familiar to the settlers. Only a few tentative and abortive early efforts thus were made from Jamestown to 'Ould Virginia,' 'Rawliana,' or 'Carolana,' the three names given in the seventeenth century by mapmaker John Farrer to the area that would become North Carolina. Farrer placed in Albemarle Sound an entry reading, 'Rolli passd,' thus helping fuel the myth that Sir Walter did indeed sail to North America. Furthermore, as late as 1651 Farrer clung to the belief that the Pacific Ocean lay just across the Appalachians.

Our earliest eyewitness account of the earthen fort left by the colonists in the 1580s dates from two years after Farrer's map, and even it is secondhand. Francis Yardley of Virginia wrote that a party led by a young trader of beavers had gone to Roanoke Island, where it came upon 'Indians a-hunting, who received them civilly, and shewed them the ruins of Sir Walter Ralegh's fort, from whence I received a sure token of their being there.' Token? How we wonder what the 'token' was, and what the fort looked like only sixty-three years after John White returned to discover it abandoned.

Just a decade after that account, King Charles II in 1663 granted to eight Lords Proprietors the vast territory from the Spanish possession northward to the Virginia border, thence westward to the Pacific. Already English settlement had begun on the mainland up the Albemarle Sound, but the alternately sandy and marshy Roanoke

Island discouraged habitation by more than a few whites, even though the island as early as 1676 had been designated for the location of the 'Chiefe towne' by the proprietors. Not surprisingly, therefore, we must wait for the next account until 1701, when John Lawson, a surveyor and later co-founder with Baron von Graffenried of the settlement of New Bern, visited Roanoke Island and reported that 'the Ruins of a fort are to be seen at this day, as well as some English Coins which have been lately found and a Brass-Gun, a Powder-Horn, and one small Quarter deck-Gun, made of Iron Staves, and hoop'd with the same Metal; which Method of making Guns might very probably be made use of in those Days, for the Convenience of the Infant-Colonies.' Lawson also repeated a local Indian legend that 'the Ship which brought the first Colonies, does often appear amongst them, under Sail, in a gallant Posture, which they call Sir Walter Raleigh's Ship.'

Throughout North Carolina's proprietary period, which ended with the purchase of the colony by the Crown in 1728, its immigrants were too busy scratching out a living in a virgin land to pay much attention to history. Five generations had passed since the stirring days of Sir Walter Raleigh and the Roanoke expeditions. Furthermore, the increase of non-English immigrants beginning in the 1740s salted the colony with other nationalities less interested in the heroic exploits of the British. All the while, Roanoke Island and the outer banks remained little more than barriers to navigation between the Albemarle Sound and the Atlantic. It is all the more puzzling, therefore, why surveyor John Collet's map of 1770 for the first time firmly planted at the northeast edge of Roanoke Island the symbol and word 'Fort' and in the waters of Roanoke Sound the words 'Rawleigh' and 'Walter.' Thus nearly two centuries had passed before a cartographer confirmed the location of the fort built by Ralph Lane's colony.

Ironically, it was the American Revolution against the Mother Country that started the rehabilitation of Sir Walter Raleigh as the symbolic founder of English America. Six years after the stamp act crisis and three years before the clash of American and British arms at Lexington and Concord, the little county of Chowan up the Albemarle Sound from Roanoke Island adopted a seal depicting Sir Walter Raleigh standing on the shore with a shield and liberty pole

75

and carrying this inscription: 'Sir Walter Raleigh landed in America A.D. 1584' (Plate 1). The myth persisted.

War summons history to its support, and the name of Sir Walter Raleigh was invoked in the cause of American independence, even against his native land. It is entirely possible that Joseph Hewes was responsible for his county's adoption of the seal just mentioned, and it is almost certain that this influential member of the Continental Congress—he is sometimes called the 'father' of the United States Navy—was responsible for the naming of a thirty-two gun, three-masted Continental frigate built in 1776 in Portsmouth, New Hampshire. After toasts to 'prosperity, freedom & independency to the thirteen united states of america,' the *Raleigh* sailed off for distinguished service before being seized off the coast of Maine by the English, who turned the vessel against the colonists and in 1781 sailed it to Portsmouth, England, only a hundred miles or so from Hayes Barton. Even after this ignominious end, a crude drawing of the *Raleigh* graced the official seal of the state of New Hampshire.

The Continental frigate was only the first of several American ships bearing the surname of the Shepherd of the Ocean. During the Civil War, the Confederate *Raleigh* operated in North Carolina's coastal waters. A third Portsmouth—this one in Virginia—was the scene of the launching in 1894 of a 305-foot, 3,100-ton armored cruiser that participated with Admiral Dewey in the capture of the Philippines. After the Spanish-American War, this cruiser was in and out of service until World War I, during which it performed valiantly before being decommissioned and scrapped. In 1922 another cruiser named *Raleigh* was launched in Quincy, Massachusetts. Badly damaged by the Japanese at Pearl Harbor, this *Raleigh* was rehabilitated and for a time served as flagship for Admiral Leahy in World War II. In 1962 a smaller *Raleigh*, this one an amphibious troop and craft carrier, went to sea.

Apparently the first official use of the name Raleigh by the legislature of North Carolina occurred shortly after the Revolution when in 1787 it chartered a Raleigh Company for the construction of Raleigh Inlet through the outer banks northeast of Roanoke Island. The inlet or 'canal' appears never to have been cut, and today only Raleigh Bay is found on coastal maps. Long gone are Ports Fernando, Grenvil, and Lane, which were found on early maps.

For more than a century of European migration, North Carolina's colonial government was peripatetic, without a permanent seat for its officials and assembly. Governor William Tryon tried to establish the capital at New Bern, but the government had to flee inland to escape capture by the British during the Revolution. Moreover, by the end of the Revolution, the center of population had moved westward, and a coastal capital was no longer acceptable. Regional jealousies delayed until 1788 the selection of Wake County, and four more years passed before the acquisition of the exact wooded tract from which a new capital city could be carved. At last, two centuries after the Roanoke colonies, North Carolina was ready to pay proper tribute to Sir Walter Raleigh: The 'permanent and unalterable seat of the government of the state of North-Carolina' would be the 'City of Raleigh.' Wrote a hopeful poet of the wilderness site:

> With a prophetic eye, we now behold,
> What some propitious hour may unfold.
> A lofty *Dome* magnificently fair,
> Ascending from the top of *Union-Square*...

The new village was laid out by William Christmas, lots were sold to benefit the state coffers, and a simple State House was ready for the reception of the General Assembly and cabinet officers in 1794. The building was enlarged and improved, but it burned in 1831 and was replaced by a larger granite capitol that was dedicated in 1840 simultaneously with the opening of the Raleigh and Gaston Railway. Thus from Joel Lane's woods grew a village that eventually became a vibrant town in the nineteenth century, proud of its role as center of governmental and cultural affairs of the state. Today it is a city of about 200,000 population, more conscious than ever of the significance of its name.

Its municipal flag, for instance, displays Sir Walter Raleigh's coat of arms; and among the communities within the city are Budleigh and Hayes Barton (there is, for example, the Hayes Barton Baptist Church); and among its streets are Devon Avenue, Devon Circle, Devon Court, Devonshire Drive, and Exeter Circle. Businesses (Sir Walter Chevrolet), services (Sir Walter Beauty Trends), professions (Sir Walter Opticians), housing complexes (Sir Walter Apartments), and organizations (Sir Walter Ski Club) are among others carrying his name. The city once had not only a Raleigh Hotel but also a Sir Walter Hotel, the latter featuring on the wall of its famous Virginia Dare Ballroom a courteous Raleigh spreading his coat over the mud to protect the feet of his queen. A luxury motel named The Velvet Cloak dates from the tercentenary of the Carolina Charter in 1963. A prestigious recognition is given by the Raleigh City Council—the Sir Walter Raleigh Award for community appearance. (A new but traditionally designed structure near the Executive Mansion received one of this year's awards. Perhaps appropriately, it is occupied by the North Carolina Beer Wholesalers Association.) Only last month the Sir Walter Raleigh Road Race drew thousands of runners to the capital city.

Raleigh's name is by no means limited to the city called after him. At least six railroads have carried it, including a locomotive captured during the American Civil War. His name and that of others associated with him are common on Roanoke Island and along the outer banks, particularly in the county of Dare and the towns of Manteo and Wanchese. North Carolina's award for the best book of fiction is the Sir Walter Raleigh Award, given by the Historical Book Club of North Carolina; and the wives—now spouses—of state legislators form the Sir Walter Cabinet. Sadly, our New Exeter along the Cape Fear River disappeared more than a century ago, and our Plymouth, North Carolina, traces its name to England only through the one in Massachusetts. The 'Exeter' of Thomas Wolfe's novel, *Look Homeward, Angel*, is in fact our city of Durham, the home of Duke University.

Nor has Raleigh's name been limited to North Carolina. There are places named Raleigh in Georgia, Illinois, Mississippi, North Dakota, and Tennessee, plus the county of Raleigh in West Virginia; and it is claimed that the town of Rolla in Missouri was suggested by an emigrant from North Carolina who could not spell. If so, the towns of Rolla in Kansas and North Dakota may have the same derivation. There are a popular Raleigh Department Store in Washington, a Raleigh Tavern in Williamsburg, Virginia, and a Raleigh Hotel in Waco, Texas. My favorite hotel in Washington surprised me in 1963 with this response to my request for a reservation: 'The Raleigh Hotel will cease to exist. It is being demolished for another government office building.' In Canada's

British Columbia there are Raleigh Passage, Raleigh Point, Raleigh Ridge, and Mount Raleigh; in Ontario, Raleigh, Raleigh Falls, Raleigh Island, and Lake Raleigh; in Newfoundland, Raleigh, a tiny community only a few miles from the remains of the earliest Norse settlement of about 1000 AD; and near my favorite Arctic escape on Baffin Island, Mount Raleigh and Raleigh Glacier. Far up the lonely Alcan Highway in Canada's Yukon Territory, I found the name Raleigh in a forest of signs.

Raleigh has not escaped the lure of name brands. In my youth a house-to-house salesman representing a firm called Rawleigh's supplied our few household needs. Raleigh has also attained aromatic immortality. In 1899 the R. J. Reynolds Tobacco Company produced Walter Raleigh Fine Golden Twist chewing tobacco; and today's Raleigh Cigarettes and Sir Walter Raleigh Smoking Tobacco—regular, special blend, and aromatic—while using Tar Heel (North Carolinian) tobacco, are made in Kentucky, where the Brown and Williamson Tobacco Company displays in its headquarters an amusing smoking statue of Raleigh. Insofar as I can determine, the only type of tobacco that has not capitalized on his name is snuff. For some reason, a popular wine was named for Virginia Dare rather than Sir Walter.

Let us return to Fort Raleigh, the spiritual birthplace of English America, for only it provided a direct physical link with the events of the 1580s. Our fifth president, James Monroe, was taken in 1819 to 'the remains of the Fort, the traces of which are still distinctly visible, which is said to have been erected by the first colony of Sir Walter Raleigh.' A decade later the historian Francois-Xavier Martin wrote, 'The stump of a live oak, said to have been the tree on which the word [CROATAN] was cut, was shown, as late as the year 1778, by the people of Roanoke Island.' Unfortunately, Martin was by then losing his sight and living in Louisiana, so he could not point out the spot.

About thirty years later an itinerant teacher visiting Nags Head, George Higby Throop, wrote that 'the remains of the fort, glass globes, containing quicksilver, and hermetically sealed and other relics occasionally discovered there, give rise to a thousand conjectures destined never to be solved.' About 1850, the traveling Benson J. Lossing reported that 'slight traces of Lane's fort' could still be seen near the north end of the island, but he left no pencil sketch such as those that he drew at other important sites.

Though he gave no description of the fort, historian John Hill Wheeler in 1851 wrote of 'the generous, chivalric and noble Raleigh':

> His memory sparkles o'er the fountain:
> His name inscribed on lofty mountain,
> The meanest rill, the mightiest river
> Rolls, mingled with his name forever.

In 1860 Edward C. Bruce, a Virginia artist, visited the site and wrote in an article in *Harper's New Monthly* Magazine,

> The trench is clearly traceable in a square of about forty yards each way. Midway of one side...another trench, perhaps flanking the gate-way, runs in some fifteen or twenty feet...And on the right of the same face of the inclosure, the corner is apparently thrown out in the form of a small bastion. The ditch is generally two feet deep, though in many places scarcely perceptible. The whole site is overgrown with pine, live-oak, vines, and a variety of other plants, high and low. A flourishing live-oak, draped with vines, stands sentinel near the centre. A fragment or two of stone or brick may be discovered in the grass, and then all is told of the existing relics of the city of Raleigh.

Even before a year had passed in the American Civil War, Union forces occupied the island, and the curiosity of history-conscious Yankee soldiers was whetted. One wrote that his favorite pastime was 'searching for traces of, and speculating upon the fate of, that lost colony...' Another wondered how the English 'ever found it or why they chose it as an abiding place,' adding, 'It is easier to account for the fact that the colony was abandoned.' Still another, Charles F. Johnson, drew a sketch of the wooded site. About this same time, Frederic Kidder, a New Englander who had lived a few years in North Carolina, wrote about his second visit to the fort site: '...the presents remains...are very slight, being merely the wreck of an embankment. This has at times been excavated by parties who hoped to find some deposit which would repay their trouble, but with little success, a vial of quicksilver being the only relic said to have been found.' Kidder was wrong; one important artifact recovered was a sixteenth-century axe, now the proud possession of the North Carolina Collection of the University of North Carolina at Chapel Hill.

During and after the Civil War there was a new surge of interest in Raleigh and his colonies. In 1862 Samuel G. Drake wrote 'A Brief Memoir of Sir Walter Raleigh' for the *New England Historical and Genealogical Register*, and the article was revised and published by Drake in a splendid pamphlet by the same title, complete with black-and-white engravings of Raleigh and men associated with him. Six years later, George Bancroft, former United States minister to Great Britain and then minister to Berlin, obtained copies of four Ralph Lane letters, and these 'entirely fresh materials' were published in the *Transactions and Collections of the American Antiquarian Society*. In 1882 seven of John White's watercolors from the British Museum were published with Edward Eggleston's article, 'The Beginning of a Nation,' in *Century* magazine; and in the same year the 'Raleigh Window,' a gift from American citizens, was dedicated in St Margaret's Church, Westminster, London.

The tercentenary of the Amadas and Barlow expedition in 1884 afforded North Carolinians with an unprecedented opportunity for countering the American myth that English America started with Jamestown or Plymouth Rock. At the urging of Richard B. Creecy of Elizabeth City, former governor and then Senator Zebulon B. Vance introduced in the United States Senate Resolution 45 calling for an appropriation of 30,000 dollars and the appointment of a joint select committee of Congress 'to prepare a design and arrange for the erection of a suitable monument or column on or near the spot where Raleigh's first expedition landed on Roanoke Island...' In his supporting speech, Vance in the flowery language of his day traced the English settlement of America, noting that 'Virginia, the virgin land of his [Raleigh's] poetic fancy, was settled, by other hands than his, it is true, but in pursuance of his designs, in consequence of his discoveries, in the track marked out by his genius, and by his own countrymen.' But for one voice the resolution appeared headed for unanimous passage in the Senate. That voice was not a Virginian's, as we might expect. Instead, it was the single voice of a senator from Kansas, who insisted that the resolution be referred to a committee where, true to Sir Walter's fate, it was buried. Upon investigation, it was found that Senator John J. Ingalls was a native of Massachusetts, the home of the writers of most of the American history books to that time.

Vance's efforts, however, stimulated extensive press coverage, and in the fall attorney and editor Samuel A. Ashe delivered before the North Carolina Press Association a lengthy oration on Sir Walter Raleigh and his attempts to colonize America, proclaiming that 'The English nation he expected to behold is here in Carolina, and the City of Raleigh is its metropolis.' That same year J. E. Goodwin of Manteo described the fort site as 'like a cluster of caves with banks thrown up between them.'

The year 1884 thus passed with little more than orations and newspaper lamentations that no official observance was being held, but the subject would be quiet no longer. Hamilton McMillan, a state legislator, four years later published a little book purporting to prove that a large group of citizens of the Robeson County area, who resisted efforts of whites to classify them as blacks, were descendants of the lost colonists, who had intermarried with Indians. A professional historian, Stephen B. Weeks, carried this theory to the pages of the *Magazine of American History* a few years later, and the myth was spawned that the mystery of the John White colony had been solved.

Virginia Dare captured the imagination of women, and in 1892 the Virginia Dare Columbian Memorial Association was organized to 'perpetuate the memory of Virginia Dare, the first white child born on American soil, to erect a memorial to her in North Carolina and to aid in the construction of a building for the State of North Carolina at the Chicago World's Columbian Exposition.' Virginia Dare was not in fact the first white child born in America (what about the Spanish in Florida?), and the chief legacy of the short-lived association was a splendidly carved 'Virginia Dare Desk' made of white holly from Roanoke Island, exhibited at Chicago, and now in the state museum of history.

It was another organization, the Roanoke Colony Memorial Association, organized by expatriot North Carolinians in Baltimore, Maryland, that succeeded in acquiring and preserving the Fort Raleigh site. Led by two brothers, confusingly named Edward Graham Daves and Major Graham Daves, the association was incorporated in 1894 'for the benevolent and patriotic purpose of reclaiming, preserving, and adorning Old Fort Raleigh, built in 1585...' Sufficient funds were raised by 1896 to acquire the property and to erect a granite marker, but not until archaeologist Talcott Williams of the University of Pennsylvania Museum had conducted sufficient excavations to satisfy Trinity College historian John Spencer Bassett that the site was indeed authentic. The marker read, in part, 'On this site, in July-August, 1585 (O.S.) colonists, sent out from England by Sir Walter Raleigh, built a fort, called by them "The New Fort in Virginia"...' Tradition was finally translated into history, and Roanoke Island in succeeding years was the scene of periodic observances featuring challenges that were ignored by the people of 'New' Virginia and Massachusetts. In 1902, for example, Chief Justice Walter Clark spoke on the subject, 'The Cradle of American Civilization': 'Here, 36 years before the landing of the Pilgrims at Plymouth Rock, here 23 years before John Smith and Jamestown, in the year 1584, the first English keel grated on the shores of what is now the United States. Here the greatest movement of the ages began, which has completed the circuit of the globe.'

The following year the General Assembly authorized creation of the Roanoke Island Celebration Company to 'promote, organize and conduct on Roanoke Island...a celebration of the landing and settlement of Sir Walter Raleigh's colonies,' to collect documents and artifacts, and to establish 'an institution for investigating and teaching useful arts and sciences...' The great exhibition was scheduled for 1905, two years before the tercentenary of the Jamestown colony. The organization failed to raise even the required amount to declare itself in business, and the movement came to naught. The ultimate insult came when the Daughters of the American Revolution proposed—unsuccessfully, I am glad to say—a fundraising campaign for the construction of a replica of Hayes Barton, not on Roanoke Island but in Jamestown, Virginia! Even so, North Carolina provided an exhibit at the Jamestown celebration, including copies of some of the John White drawings (paid for by Colonel Bennehan Cameron) and thirteen original paintings by Jacques Busbee depicting the Fort Raleigh area in 1906.

The new century brought a reawakening of North Carolinians to their backwardness in saving and promoting knowledge of their history, and Raleigh and his colonies occupied about fifty pages in the compendium, *Literary and Historical Activities in North Caro-*

lina, 1900–1905, including Henry James Stockard's poem, 'Sir
Walter Raleigh,' with these lines:

> He is not greatest who with pick and spade
> Makes excavations for some splendid fane [temple];
> Nor he who lays with trowel, plumb, and line
> Upon the eternal rock its base of stone...
>
> These are but delvers, masons, artisans,
> Each working out his part of that vast plan
> Projected in the master builder's brain.

At the end of a sixteen-page chronology of Raleigh's life, William J.
Peele added.

> His name and fame are indissolubly linked with North Carolina. He
> made the first chapter of her history, which is also the first chapter of
> Anglo-American history, and one day the English-speaking race on this
> continent, with the Carolinians in the lead, will call its brethren across
> the seas and go back to the island where it began its conquering march
> to do honor to the man who gave himself and all he had for its
> advancement.

Periodic commemorations conducted at or near Fort Raleigh by
the Roanoke Colony Memorial Association continued both before
and after World War I, but ceremonies of a different kind were
proposed in 1918, the three-hundredth anniversary of Raleigh's
beheading. The idea originated with Dr Israel Gollancz, secretary of
the British Academy, who after helping organize a Raleigh
Tercentenary Committee in London, approached the American
ambassador, Walter Hines Page, a native North Carolinian, who put
him in touch with the State Literary and Historical Association. This
was to be a scholarly conference, as these topics indicate: 'Ralegh and
British Imperialism,' by Edwin Greenlaw of the University of
North Carolina; 'Sir Walter Raleigh as a Man of Letters,' by Frank
Wilson Chaney Hersey of Harvard; 'Raleigh's Place in American
Colonization,' by Charles M. Andrews of Yale; 'England and the
Birth of the American Nation,' by William Thomas Laprade of
Trinity; 'The Converging of Democracies of England and America,'
by William E. Dodd of the University of Chicago; 'Anglo-American
Diplomatic Relations during the Last Half-Century,' by Charles H.

Levermore of the New York Place (*sic*) Society; and 'Social and Political Ideals of the English-Speaking Peoples,' by George A. Wauchope of the University of South Carolina. But, alas, the conference was cancelled because of the Spanish influenza pandemic that swept the world, killing an estimated twenty million people, including more than 13,000 North Carolinians. Wisely, however, the association persuaded the invited speakers to submit their papers, which were published the following year. Nevertheless, on October 29, 1918, the date of the planned conference, greetings were exchanged by cable, and Earl Fortescue, lord lieutenant of Devon, sent this message from Exeter: 'Devonshire sends greetings on occasion of Raleigh Tercentenary.'

Official recognition was finally given Fort Raleigh in 1926 by the governments of Great Britain and the United States. Ambassador Sir Esme Howard was the speaker at ceremonies dedicating two plaques made possible by a 2,500 dollar congressional appropriation. Significantly also, *The Times* of London carried a lengthy story on the event and its meaning.

As early as 1920 drama was introduced as a medium for telling the story of the Roanoke voyages. *Raleigh, the Shepherd of the Ocean,* written by Professor Frederick Koch of Chapel Hill and directed by Elizabeth Grimball of the New York School of the Theatre, was presented in Raleigh three nights during the state fair. The following year Miss Grimball was hired by the State Board of Education to direct one of the earliest motion pictures made in the state. This first *Lost Colony,* the script of which was written by Mabel Evans, was shot on location near Fort Raleigh, with North Carolina actors. The five-reel silent film with captions was shown over the state and as far away as Boston. For many Tar Heels, it was their first 'moving picture.' Dramatic scenes continued to be played out at occasional Virginia Dare celebrations, the most impressive one being in 1934 on the 350th anniversary of the Amadas and Barlowe expeditions, when a commercial company put on for three nights *O Brave New World,* or *A Pageant of Roanoke.* This was, of course, a far cry from the great commemoration proposed in 1932 in Congressman Lindsay C. Warren's ill-fated congressional resolution. The Depression era did, however, bring activity to Fort Raleigh, for several log structures were put up by federal agencies, much to the distress of historians.

I am glad to report that they detriorated rapidly and were eventually removed.

Meanwhile, a new organization, the Roanoke Island Historical Association, was formed in 1932, and it explored the idea of an outdoor play near the fort site. From this inspiration came Paul Green's pioneering outdoor symphonic drama, *The Lost Colony*, which premiered in 1937 and was an immediate 'hit'. The eyes of the nation were on Roanoke Island when, on the night of Virginia Dare's birthday in 1937, President Franklin D. Roosevelt watched the drama and signed a delayed birth certificate—350 years late—for the first child born of English parents in America. Simultaneously the Treasury Department issued a special commemorative half-dollar and the Post Office Department issued a commemorative stamp in honor of Virginia Dare.

At last world attention was drawn to the 350-year-story of the Roanoke colonies. Already the Roanoke Colony Memorial Association had deeded title to the site to the North Carolinia Historical Commission in 1934, and five years later the site was accepted by the United States Department of the Interior, which proceeded to incorporate it and surrounding lands into the national park system following scholarly investigations by archaeologist J. C. Harrington. The earthworks were painstakingly reconstructed, the grounds marked, a visitor center museum constructed for the display of recovered artifacts and presentation of audiovisual programs, and a 'living history' program established. As developed over the years, the Fort Raleigh National Historic Site is adjacent to the Waterside Theatre, where *The Lost Colony* is performed nightly during the summer, and to the Elizabethan Gardens, owned by the Garden Club of North Carolina. Fort Raleigh, combined with the Wright Memorial where powered flight began in 1903, and the Cape Hatteras National Seashore, provides historical diversions for the waves of tourists who surge to the nearby beaches.

So the site had been saved, but what about Sir Walter himself? Well, Sir Walter wasn't forgotten. For example, when the Roanoke Colony Memorial Association disbanded, it left its modest treasury to the North Carolina Collection at Chapel Hill for the development of a Sir Walter Raleigh Collection which has, over the years, grown into the finest accumulation of Raleighana in America, with its own

Plate 1. Colonial seal of Chowan County, North Carolina, 1772, erroneously depicting Sir Walter Raleigh landing in America in 1584. Photograph by courtesy N.C. Department of Archives and History.

Plate 2. The launching of the *Elizabeth II* at Manteo, Roanoke Island, North Carolina, on 22 November 1983. Photograph by courtesy of Elizabeth II State Historic Site Visitor Centre.

bookplate (designed by artist Clare Leighton) and special rooms (containing paneling from a house not far from Exeter) and furnishings. Of our three statues of Raleigh, the eighteenth-century wood likeness, attributed to a student of Grinling Gibbons, is in my opinion surpassed by none in the world.

Still, North Carolina was without a reputable outdoor statue of Sir Walter. True, an ambitious movement was launched in 1901 to provide a statue for Nash Square in the capital city, but even the talents of a leading citizen, Julian S. Carr, failed to raise the proposed 100,000 dollars for a mighty figure of Raleigh, surrounded by lesser figures of Thomas Harriot, representing science; John White, representing art; Richard Hakluyt, representing history; and Ralph Lane, representing exploration. Periodically the drive was renewed, but the fund, including thousands of pennies donated by school children, was partially lost during bank failures in the Depression. North Carolinians could console themselves, however, by pointing out that even London did not have a public statue of Raleigh until 1959, when the one in Whitehall was made possible by funds generated by a group headed by John Bigelow Dodge, whose wife just happened to be Minerva Arrington, a native North Carolinian. When Governor Luther Hodges, after visiting and approving the London statue, proposed that sculptor William McMillen produce for North Carolina a copy of the Whitehall piece, he stirred up a tempest: 'No Copycat Carolina,' headed an editorial in the *News and Observer*. Seventeen more years passed before North Carolinians could see an outdoor statue of Sir Walter Raleigh in the city named for him— three-quarters of a century after the first pennies were collected.

Paul Green's *The Lost Colony* has for forty-eight years (with the exception of three war years) kept the Roanoke story before Americans each summer in the Waterside Theatre at Fort Raleigh. Hundreds of thousands have visited the site and watched the outdoor symphonic pageant, which, incidentally, was the pioneer in a new form of drama that has spread across the country. Periodically, 'Sir Walter Raleigh Day' has been observed in our public schools. Of enormous influence was the exhibition in 1965 of original John White drawings coinciding with the publication of the two-volume, limited-edition *The American Drawings of John White*, by Paul Hulton and David Beers Quinn, a joint project of the University of

North Carolina Press and the British Museum. Periodically, also, 'Elizabethan Festivals' have been held, and there have been other reminders of Raleigh and his colonies.

But the impetus for the present quadricentennial commemoration came in 1955 from a former congressman and comptroller general of the United States, Lindsay C. Warren, Sr, who persuaded the General Assembly to authorize the governor to appoint 'The Commission to Promote Plans for the Celebration of the Four-Hundredth Anniversary of the Landing of Sir Walter Raleigh's Colony on Roanoke Island', with a charge to observe the occasion by 'a National or World's Fair Exposition, or in some other suitable manner...'. After holding just one meeting the commission faded into obscurity, perhaps because the proposal seemed to emphasize entertainment over history, imagination over fact, and a 'colossal celebration' over mature commemoration. At least, that was my reaction when, as director of the North Carolina State Department of Archives and History in the early seventies, I began working with Governor Robert W. Scott in drafting legislation for the consolidation of all cultural agencies into the Department of Cultural Resources. I therefore redrafted the purposes of the resolution into a statute that envisioned a history-oriented four-year observance.

At that time, I had no way of knowing that in 1978, after I had moved to the University of North Carolina, Governor James B. Hunt, Jr, would appoint me chairman of America's Four Hundredth Anniversary Committee and give me the unprecedented privilege of selecting the other nine appointed members, whom he duly commissioned. It was this group who for two years carefully laid plans for a multifaceted program, the core of which would be scholarly research and publications legitimizing our claims of the importance of the events being commemorated. We Americans tend to frivolize—if I may coin a word—important historical events, and I was determined that this would be an exception. Our foremost objective was to be sure that the American people would in 1987 know that English America traces its roots to Roanoke Island.

At the core of this plan was scholarly historical and archaeological research, culminating in a three-level series of leaflets, booklets, and major books, some of which have already appeared, and others of which are in preparation. These, we knew, would spawn private works, several of which have now been published. Thus we sought to put the Raleigh story into both professional and popular readership. That objective taken care of, we then turned to plans for exhibitions and cultural activities (including an opera by Ian Hamilton, art and literary competitions, workshops, lectures, television specials, etc.); public ceremonies that were marvelously opened on Roanoke Island by Princess Anne in July 1984; and the building of the *Elizabeth II*, christened by the princess and featured on a special commemorative stamp issued by the Postal Service. This ship, of course, draws far more attention than the other efforts of the commemorative committee, and that is why I and my successors insisted on as authentic a 'representation' of a sixteenth-century sailing vessel as is possible without detailed drawings. We have been pleased with the result (Plate 2).

There have been, and will be, many more activities during the era ending in 1987, but my own pride is greatest in the British Library exhibit on display in the North Carolina Department of Archives and History in Raleigh. Dr Helen Wallis and her staff put on the 'Raleigh & Roanoke' exhibition first in the British Library in 1984, then brought it to America for opening in Raleigh on March 6, 1985. In June it was moved to the New York Public Library. Every American ought to see it. The catalog, written by Dr Wallis, is in itself a major contribution to America's quadricentennial. Senator Lindsay C. Warren, Jr, the new chairman, and his committee have done not only North Carolina but all of America proud. Visit North Carolina during the quadricentennial, which has more than two years still to go.

Early on, I mentioned the ubiquitous myth that Sir Walter Raleigh came to North America. We thought we had made headway in stamping out this mistaken notion, but, alas, when a new history textbook was adopted two years ago for use in our public schools, there on page 76 appeared a drawing with this caption: 'Sir Walter Raleigh directs the raising of the English flag on the coast of Virginia.' There was, of course, an immediate outcry, and the error was quickly corrected in subsequent printings with the substitution of a picture of Sir Walter and his son Wat.

The story of Sir Walter Raleigh, even if a little distorted, is alive and well in North America.

6

The Raleghs, Father and Son

1. *Walter Ralegh 1505–1581*

Michael J. G. Stanford

Much has been made of the piratical and privateering aspects of the Roanoke voyages. I do not dissent from that view: indeed I believe it is supported by the family and social background of Sir Walter Raleigh. A greater knowledge of his immediate family—parents, elder siblings, uncles and aunts—can also throw light on the enigma of Sir Walter himself and on that west country society of small gentry and thrusting merchants from which he came. A French historian, Pierre Lefranc, writes that when Sir Walter was at the peak of his success as a courtier (my translation),

> he kept in close relations with the milieu from which he sprang, and continued to challenge the power of Spain in company with the sailors and gentlemen of the West Country.[1]

What was the nature of this West Country community which supplied at least half of England's leaders in the long sea war with Spain? Was it a class, or a clan, or a status group? How united was it in interest, in ideas and in actions? The conventional view, maintained at least since the appearance of Froude's *History of*

91

England in the late 1850s and 1860s, is that the ethos of these westcountrymen was one of maritime enterprise, of Protestantism and of fierce antipathy to Spain. I wonder how far nineteenth century enthusiasms are still being attributed to men and women who lived four centuries ago.

The Ralegh family was certainly old and honourable, though not particularly distinguished. It can be traced, perhaps from the early twelfth century, and with fair certainty from Sir Wymund Ralegh who died in 1258 and who already held some of the lands that came down to Sir Walter and his family.[2] But we can begin our story with a later Wymund Ralegh, son of a Walter Ralegh and a Katherine Champernowne. He lived from 1475 to 1512, roughly the reign of Henry VII.[3] Wymund was the ward of Sir Richard Edgcumbe of Cotehele and married Sir Richard's daughter, Elizabeth. In 1502 the Star Chamber fined him 700 marks.[4] His offence is unknown but it may have been connected with the Cornish rebellion five years earlier. To meet this Wymund made over the family estate of Smallridge in east Devon, which the family had held for nearly three hundred years, to Sir John Gilbert, and his other four manors to his brother-in-law, Sir Piers Edgcumbe.[5] Edgcumbe, and three other gentlemen, then paid Ralegh's fine, recovered the money from the income of the manors by 1507 and then retained them in trust for Wymund and his heirs, the eldest of whom, Walter, was born in 1505.[6]

Seven years later Wymund died, leaving another minor liable to the exactions of wardship. Edgcumbe and his associates tried to conceal the matter but the facts came out and the wardship, with the four manors, went, not to a friendly neighbour but to a stranger, Sir Nicholas Vaux.[7] The ending of the wardship in 1526 was a turning point in the history of the family. Vaux released the lands to Piers Edgcumbe and Walter Ralegh, but Ralegh himself chose to live, not on any of his ancestral estates, but at Hayes in East Budleigh.[8]

In, or perhaps before, 1526 Walter married Joan, the daughter of John Drake of Ash in the parish of Musbury.[9] These Drakes were small landed gentry, perhaps a little lower in the social scale than the Raleghs, and several of them were doing well as merchants both in Exeter and in Exmouth. It can be assumed that it was the Drake marriage which led the Raleghs to seek, in maritime activities, a means of recovering from the financial blows of two wardships and the Star Chamber fine.

The Drake marriage must have occurred no later than 1526 because in 1527 was born George, the eldest of Walter's sons, the name, incidentally, being new to the Ralegh family. A year or so later a second son, John, was born.[10] In 1530 Walter was widowed and his wife's tomb may still be seen in East Budleigh church. He married again and had a daughter, Mary, though who his second wife was is still not entirely clear.[11] It was nearly two decades later, between 1547 and 1549, that Walter married yet again, this time Katherine, née Champernowne, the widow of Otho or Otes Gilbert.[12] She was already the mother of three sons, John, Humphrey and Adrian. Why she should marry Walter Ralegh, a man older than herself and her social inferior, is not clear. Was it love? In the Tudor period, however, sons seem to have married better, in the social sense, than daughters. The Ralegh boys married Champernowne (c. 1470), Edgcumbe, Drake, Gilbert, Fortescue, Champernowne again, Wroughton and Throgmorton. The daughters married Courtenay, Upton, Snedall, Radford and Hall. Although a small sample this seems to bear out Lawrence Stone's findings for the Tudor aristocracy.[13]

In 1550 Katherine bore a fourth son, Carew Ralegh, and in 1552 or 1554 a fifth, who was given his father's name, Walter. She also had a daughter, Margaret, and if Margaret came between Carew and Walter, as one genealogy suggests, then 1554 is to be preferred as the date of Walter's birth. In 1551 Walter senior renewed the lease of what was later to be known as Hayes Barton for himself and his second son, John. In 1560 he made some small financial provision for his two youngest sons, Carew and Walter, and then, sometime in the late 1560s, he moved to Exeter.[14] He was by then over 60 years old, a good age, by Elizabethan standards, for retirement.

Much of the evidence shows Walter Ralegh senior pursuing at Hayes the usual activities of a landed gentleman. He engaged in law suits and performed a round of the public duties which Tudor governments imposed upon men of his class. It is not surprising that in 1543 he was called upon to raise men for the French wars, but less typical that in 1545 he was required to stay in Devon should the levies be called upon to serve outside the shire.[15] In 1558 he was again an officer, this time for coastal defence.[16] In this year he sat in Mary Tudor's last parliament as a member for Wareham in Dorset, and he made use of this almost at once to claim parliamentary privilege against a summons by the High Court of Admiralty.[17]

So far this is a fairly routine life, but there are two other areas of some interest: religion and the sea. Everyone knows the story told by John Hooker, of how in 1549, during the Prayer Book Rebellion, Walter berated an old woman for telling her beads and so brought on himself the whole congregation of Clyst St Mary, so that

> if he had not shifted himseffe into the chaple there and had [not] benne rescued by certeyne Maryners of Exmouthe which came with him, he had benne in greate daunger of his lieffe and lyke to have benne murdred.[18]

After this he was captured and imprisoned in St Sidwell's church for the duration of the siege of the city of Exeter.[19] On his release he made off with some of the church valuables, as did others, but Ralegh, unlike his fellow prisoners, restored what he had taken.[20] Three years earlier he had been in conflict with the churchwardens of East Budleigh over a silver cross which they and he both claimed. This may be an example of Protestant looting of church goods, but we cannot tell. In 1551 Ralegh bought a bell from Sidmouth church, but quite legally. Another reason for accounting him a leading Protestant is that he helped Sir Peter Carew escape after the failure of the 1554 rising, but this may well have been for friendship rather than for religion.[21] All in all there is little evidence that Walter, as distinct from his wife,[22] was the keen Protestant he is generally held to have been.

Much the most interesting material on the activities of Walter, his sons George and John, together with the Drakes and other kinsmen by marriage, is to be found in the records of the High Court of Admiralty. Walter was the owner of several ships engaged both in lawful trade and in privateering between 1549 and 1558, though how much of either there is no means of knowing. The Exeter customs records are incomplete and the Raleghs only appeared before the Admiralty court when they had exceeded the bounds of their privateering licences and, by attacking neutral shipping, were technically pirates. How many French ships they took we have no means of knowing. They appear in court only in connection with the capture of Flemish, Portuguese, German and Scottish ships and only once does a Spanish ship appear among their victims, in spite of the supposed anti-Spanish fervour of the family.

During the last three years of Mary Tudor's reign Ralegh was acting Vice-Admiral of Devon. He was thus an officer of the Admiralty while at the same time a frequent defendant in its court. This did not embarrass him: the situation was not unusual among West Country men at the time. In one case he put up recognisances of £2000, while his brother-in-law, Sir Arthur Champernowne, stood surety for a further £2000. It seems that the family was making good its losses.

There are a number of mysteries about Walter's career. Although he found much public employment under Mary Tudor he seems to have lost all his offices after Elizabeth's accession. This seems odd if he was so keen a Protestant. Then, again, although there is no evidence of his actually going to sea, in April 1558 he was named, along with seven other 'gentlemen of the West', to take command of one of the Queen's ships at Portsmouth. Tantalisingly, we know no more. One can only surmise that Mary had to accept the services of lesser men who were indifferent in their religious allegiance while Elizabeth had a wider choice when greater men returned from exile or were keen to serve her. On the whole, however, the same family names crop up whoever was on the throne in both lawful and unlawful business in the South West.[23]

Walter senior died in 1581, John in 1588, Katherine in 1594 and George in 1597, leaving what remained of the family property in the hands of Carew and Walter junior. More important, we can be sure, was the intangible inheritance that Sir Walter took from his family and its Devonshire connections.[24]

NOTES

1. Pierre Lefranc, *Sir Walter Ralegh, Écrivain*, Paris, 1968, p. 175.
2. *Book of Fees...Testa de Nevill*, pp. 399, 791; Edward Edwards, *Life of Sir Walter Raleigh*, 1868, vol. 1, p. 6; T. N. Brushfield, 'Raleghana (iii)', *Devonshire Association Transactions*, xxxii, 1900 and John Prince, *Worthies of Devon*, 1810, p. 80.
3. British Library, Harl. MS 1080 (The Visitation of Devon, 1620); Public

Record Office, Chancery, Inquisitions post mortem, Series II, 25/36 (Walter Ralegh, ob. 1485), 30/45 (Wymund Ralegh, ob. 1512).

4. W. Campbell (ed.), *Materials for a History of the Reign of Henry VII*, 1873, vol. ii, p. 78; BL, Lansdowne MS 160, fo. 320.

5. Inq. p. m. Wymund Ralegh, n. 3 above; PRO, Court of Common Pleas, Plea Rolls, 362, m. 353 and 370, m. 127; Chancery, Close Rolls, Henry VII, 366, m. 30.

6. Inq. p.m. Wymund Ralegh, no. 3 above; *Letters and Papers, Foreign and Domestic, of the reign of Henry VIII*, 1 (i), pp. 252, 438 (3) and IV (ii), g. 2673 (18).

7. Inq. p.m. Wymund Ralegh, n. 3 above; PRO, Chancery Inq. p.m., Series II, 29/60; PRO, Chancery Warrants, 434, 467; PRO, Chancery, Patent Rolls, 627, m. 25; and PRO, Exchequer, King's Remembrancer, Memoranda Rolls, 294, m. 17v.

8. See notes 6 and 5 above.

9. BL, Add. MS 14315, fo. 35v; J. L. Vivian (ed.), *The Visitations of the County of Devon*, Exeter, 1895, p. 639; F. T. Colby (ed.), *The Visitation of the county of Devon in the year 1564*, 1881, p. 180.

10. PRO, Chancery, Inq. p.m., Series II, 194/2; Vivian and Colby, above, note 9; PRO, Court of Common Pleas, Feet of Fines, Series II, file 7, no. 36, m. 52.

11. Vivian, *op. cit.*, p. 639. But see also J. Roberts, 'The Second Marriage of Walter Rawley', *Devon and Cornwall Notes and Queries*, XXXIV, i. 1978, pp. 11–13, which shows fairly convincingly that Walter's second wife was Elizabeth, daughter of Giacoma de Ponte, a merchant of Genoa who had become a denizen in 1507.

12. Otes died in 1547: *Calendar of Fine Rolls 1547–53*, p. 319.

13. Vivian and Colby, *op. cit.* The fifteenth century Champernowne marriage may be an error: see T. N. Brushfield, 'Notes on the Ralegh Family', *DAT*, vol. XV, 1883, pp. 176–8. L. Stone, *The Crisis of the Aristocracy*, abridged ed. 1967, p. 80.

14. Brushfield, *DAT* vol. XXVIII, 1896, pp. 274–6; PRO, Chancery, Inq. p.m., Series II, 194, 2; H. G. J. Clements, 'Local vestiges of Sir Walter Raleigh', *DAT*, VI, 1873, p. 227; and PRO, State Papers Domestic Elizabeth, lvii, 2.

15. PRO, State Papers Henry VIII, 184, fo. 85 and 185, fo. 58a; State Papers Papers Domestic Edward VI, XVIII, fos 81a, 82b, no date but from internal evidence belongs to the French war of 1543–6 rather than that of 1549–50. Walter's name is absent from a similar list of 1548: S.P. Dom. Edw. VI, IV, fo. 32b.

16. PRO, State Papers Domestic Mary, vol. XII, no. 67 (18 April 1558).

17. *Return of Members of Parliament*, part I, 1878, p. 396; *Journals of the House of Commons from 1547*, index by T. Vardon and T. E. May, 1852, I, pp. 47–8.

18. John Vowell alias Hooker, *Description of the Citie of Excester*, ed. W. J. Harte *et al.*, Exeter 1919, pp. 62–3.

19. *ibid.*, pp. 65–8.
20. Beatrix F. Cresswell, *The Edwardian Inventories for the City and County of Exeter*, 1916, pp. 76–82.
21. PRO, Early Chancery Proceedings, file 1172, no. 9; Exchequer, King's Remembrancer, Inventories of Church Goods, Devon, no. 2/7, fo. 10v; and PRO, State Papers Domestic Mary, vol. III, 10, fo. 34.
22. *Actes and Monumentes of John Foxe*, ed. G. Townsend and S. R. Cattley, 1837–41, vol. viii, p. 501; T. N. Brushfield, 'Raleghana', *DAT*, vol. xxviii, 1896, pp. 280–3.
23. For an elaboration of this and the two preceding paragraphs see my article, 'The Raleghs take to the Sea', *Mariner's Mirror*, vol. 48, no. 1, February 1962.
24. For Walter senior see PRO, Chancery Inq. p.m., Series II, file 194, no. 2. An abstract of John's will, now in the PRO, appeared in *D.&C. N.&Q.*, vol. VIII, pp. 181–3. Katherine's will, which was destroyed by enemy action in World War II, was printed by T. N. Brushfield in DAT, vol. XXVIII, 1896. For George Ralegh see PRO, Chancery Inq. p.m., Series II, file 251, no. 165.

Most of the foregoing will be found in the author's MA dissertation, University of London, 1955, entitled 'A History of the Ralegh family of Fardel and Budleigh in the early Tudor period'.

2. Enigma Variations: Sir Walter Raleigh, the man and his age

Maurice W. Turner

In Elgar's 'Enigma Variations' there was the enigma theme and the variations, mysteriously delineating his friends, pictured within, as well as his own self-justification as a creative genius. Raleigh is most typically an Elizabethan enigma. Can the contradictions, vagaries and ambiguities of his character be seen as a coherent life in the context of the complexities of his time?

THE HIDDEN YEARS: 1569–74

Even for an Elizabethan fifteen years is a tender age for maturing in the terrible university of life offered in the French Wars of Religion. Raleigh makes scattered allusions to his experience in his *History of the World*, written at the other end of his life, vivid and tantalising, searingly remembered yet fragmentary. He recalls notably the mode of warfare and gives his impressions of the Huguenot leaders such as Gaspard de Coligny, whose own ambitions for French colonization may have excited the young Devonian. His subsequent experience in Ireland enriched his descriptive powers when describing the battles of the ancient world and in his *History*, commenced when nearly sixty, his most telling impression of his early years as a soldier is that it was really civil war and 'by it no nation is bettered'. In France, too, he seems to have learned a ruthlessness which was further nurtured in Ireland and which was only purged from him in old age and after much suffering. Certainly when he returned to England, to London, to the Middle Temple and to be a hanger-on at Court, he was as boisterous and aggressive as any young blade. What must have been deeply-formative years, especially in France, are lacking in detail and personal incident. One would give much to know more. Did his contribution to the Huguenot cause include privateering out of the port of La Rochelle? Otherwise whence came his maritime experience before 1578?

THE MYSTERIOUS EXPEDITION OF 1578–9

Three years after the French campaigning he was given command of a leaking 100 ton ship, one of a considerable fleet under the overall command of his half brother Humphrey Gilbert.[1] No-one knows what the original plans were but the Spanish ambassador, with an English spy on one of the ships, reported to Philip of Spain that the destination was suspected to be the island of 'Santa Genela'.[2] We know that Simon Fernandes, the Portuguese navigator, was Raleigh's master mariner on the *Falcon*. Certainly the fleet was so heavily armed with guns and soldiers as to suggest conquest rather than colonization. Was the island of Saint Helena in the south Atlantic the secret destination? The strategy of capturing islands as bases for privateering had both precedent and subsequent imitation. Or were they bound for the coast of 'Virginia' which the Portuguese navigator had already reconnoitered? By the time the ships were assembled it was too late for sailing across the Atlantic and, as Elizabethan commanders were wont to do, Gilbert and his second-in-command, Henry Knollys, fell out. Knollys sailed off to engage in piracy and storms and desertions ruined Gilbert's plans.

Raleigh alone sailed south in the *Falcon*, reportedly as far as Cape Verde or the Canaries, remaining at sea nearly six months longer than the others. The Spanish ambassador complained to the Queen of a raid on Galicia in northern Spain.[3] Who carried out such a raid? John Hooker, Raleigh's biographer, has a guarded yet appreciative comment on his hero's first known maritime adventure:

> Infinite commodities in sundry respects would have ensued from that voyage if the fleet had, according to appointment, followed you, or you yourself had escaped the dangerous sea fight wherein many of your company was slain and your ships therewith sore battered and disabled.

Within two years of these events Philip of Spain annexed Portugal, with all her possessions, including the islands.

THE COURT FAVOURITE

Although forbidding further piracy the Queen may have taken note of Raleigh's daring use of the *Falcon*. Or did her great favourite, the earl of Leicester, feel secure enough to introduce a rival to the lesser competitors, little realizing the calibre of the swaggering west-countryman? Or was it the work of Kate Ashley, his aunt, she who as governess of the Princess Elizabeth had won the everlasting affection, in trial and danger, of her royal protegée? She had introduced her other nephew, Humphrey Gilbert, as page to the Princess.[4] Raleigh's own despatches, disloyal to his superiors in Ireland, yet brilliantly descriptive of the Irish situation, may have triggered the Queen's interest at an earlier stage, and even her later choice of a nickname, her 'Oracle'. She was to delight in his cascading vision and many-facetted conversation, yet she never made him a privy councillor nor a Knight of the Garter. Or was it, in Fuller's words, the spreading of his 'new plush coat in a plashy place' which led him to become her 'dear minion' and her 'Water', mimicking his Devon accent? Perhaps it was a combination of all of these that brought him his early success.

RALEIGH'S POETRY: SINCERE OR ARTFUL?

Edward Lucie Smith writes of the Elizabethan poets that they had 'a great gift for immediacy... building poems out of whatever came to hand [and] a marvellous and noble command of imagery'.[5] He also points to their enthusiasm for translating foreign masterpieces. Raleigh's poetry, I believe, shares these characteristics but also escapes into a more unique and personal expression. There is, it is true, the poetry of the cultured gentleman, the sharing with contemporaries of an indifference to publication and a justification in the cultured habit and manner of the time. There is also the poetry composed in the intense rivalry to find favour with the Queen. Raleigh sought to match in length Spenser's epic 'Faerie Queen' in his own 'Ocean to Cynthia'. Both poets ran out of time rather than of creative intention. But the passionate courtier's devotion to Elizabeth also had an obsessional dimension, showing a deeper emotion than that of the suitor who has to be outrageously flattering and ardent to catch the special attention of the psychologically-disturbed and aging coquette. It fascinated, compromised, debased and regressed him: 'twelve years entire I have wasted in this war' is his revealing and despairing comment, suggesting several layers of conflict. His poetry of personal stress has been said to have links with Gascoigne and Wyatt and to anticipate Donne and the other seventeenth-century metaphysical poets, but it seems almost unique in its bitter cynicism and melancholia, even in an age obsessed with that particular 'humour'. The apparently Roman Catholic imagery of the Passionate Pilgrim poem is one layer in an image-laden, personal heart-cry for the fusion of warring elements within himself, and for the convergence of human and divine justice which has affinities with the Book of Job. Agnes Latham calls it 'his most naked poem'.[6] Finally there is the love poetry for his wife, showing his ability to create a rich, deep and lasting relationship with her through appalling dangers and distresses. His love was warmly reciprocated by Bess, who cherished his memory and his embalmed head all twenty-nine years of her widowhood.

HIS FEIGNED(?) ATTEMPT AT SUICIDE

On 27 July 1603, a prisoner in the Tower, accused of complicity in Spanish-inspired plots against King James, his letters to his wife suggest genuine depression and fears for her future well-being and that of his son Wat.[7] Was the suicide attempt an unconscious appeal for help against cruel fortune, a common mask of the suicidal in our own age? Was it play-acting to gain sympathy as he had sought it in Elizabeth's time? Robert Cecil, his supposed friend, thought it was a sham. It is possible that it was a deliberate attempt at suicide to save his property from confiscation, a direct imitation of the father of his friend the ninth earl of Northumberland, who had shot himself in 1585 before being attainted for treason and so escaped confiscation. In Raleigh's case only a table knife was to hand and so the attempt was thwarted by a rib when a dagger might have succeeded. Nevertheless the attempt, whatever its purpose, appears to have been cathartic, giving him, as is suggested by his keeper's report, emotional release.

Raleigh was full of contradictions, profound insecurity contrasted with 'damnable pride'. He seemed to display a vein of hysteria and his emotions sometimes almost overwhelmed his reason. 'My brains are broken', he exclaimed in a letter to his wife conveying the news of the catastrophe of his last Guiana voyage, with the death of his son Wat. It was said of him that he was Fortune's tennis ball and that she volleyed him hither and thither in triumph and disaster. His letters reveal almost a persecution complex in his not-infrequent complaints and appeals, yet also commendations and condolences of great compassion. His adopted mottoes are intensely personal and revealing: 'Let valour end my life', 'I neither seek death nor flee the end', 'As much soldier as an artist' and 'By love and valour'. They reveal the classic tension of the medieval knight, between honour and passion, asceticism and romance, between doing and being, a dichotomy and an ambivalence in the soul. One of the most psychologically-revealing statements in his *History of the World* shows his penchant for aphorisms or maxims:

> In all panic terrors, as they are called, whereof there is no cause known, or no cause answerable to the greatness of the sudden consternation, it is a good remedy to do somewhat quite contrary to that which the danger would require, were it such as men have fashioned in it, their amazed conceits.

THE CAGED HAWK WHOM TIME HAD SURPRISED

In 1608 Raleigh gained the immense asset of a fascinated and hero-worshipping friend, the fourteen-year-old Henry, Prince of Wales, who, more and more alienated from his father, was appalled that 'only my father would keep such a bird in a cage'. The Prince found the aging 'hawk' a thrilling substitute for his father's court, which would now be called 'gay', and he seemed to fulfill all the old man's hopes of fortune and destiny. Perhaps now there was to be fame and fortune as elder statesman and adviser to the future king: it was the scheming obsession that had characterised all his adult life. But it was also a genuine meeting of kindred spirits, the impressionable and enquiring youth and the still soaring intellect and enthusiasm of a young-in-spirit and many-sided genius. They conversed on many subjects: chivalry, hunting, ship design, naval and military strategy, creating a library, historical perspectives, colonization, Protestantism, the patronage of the arts and sciences, and many more.[8] The caged bird began to fulfil a long-cherished dream of recording his version of 'the story of all ages past' in affection and regard for the young Prince. In 1612 the first five parts of the opening volume of this incredibly ambitious work were presented to Henry. But Raleigh's hopes were dashed by the sudden death of the Prince when only eighteen years old. He was released from the Tower but the king did not inform him that he was to be played into the hands of Spain for his own destruction. He himself had, many years before, described the intoxication: 'To seek new worlds for gold, for praise, for glory'. Aged sixty-two, yet still eager for the glory and fame dangled before him, he risked all in one last challenge to Destiny, as his vessel was fatefully named, but his abject failure and return sealed his fate.

THE FINAL ACT AND APOTHEOSIS

Time had run out on him: Destiny called him to the scaffold. Could he find transcendence in what he had called 'this stage-play world'? His enemies called him 'hypocrite', meaning 'play-actor'.

He was a true child of his time with its awareness of the stalking shadow of death and the medieval concept of making a good end. He also possessed the Elizabethan sense of the drama of life. The stage was set: Sir Walter's compulsion was ever to outdo, a kind of

Platano. *or Planten.*

Plate 3. John White's watercolour drawing of Banana Fruits. Copyright:
Trustees of the British Museum.

The manner of their attire and
painting them selues when
they goe to their generall
huntings or at theire
Solemne feasts.

Plate 4. John White's watercolour drawing of an Indian Chief painted for
the hunt. Copyright: Trustees of the British Museum.

Spenserian 'magnificence'. He had carefully prepared and rehearsed his long apologia speech the night before, when all other men slept. There was the inter-play with his old servant Peter in the matter of the cap, the invitation to friends among the nobility to stand closer that they might hear the better, and his conversation with the sheriff and the Dean of Westminster. He verbally fenced with the executioner, practitioner of the 'sharp medicine to cure all his ills'. He chose the way his head should lie on the block. Finally he climaxed his stage-managing of the whole scaffold scene by prompting the executioner, dictating when the axe should fall: 'What dost thou fear? Strike, man, strike!' And overnight he and his 'play of passion' became a national treasure.[9]

But it was not merely a charade he sought, in his conscious and very courageous acting out of the drama on the scaffold. It was all this, but heaven too. In his bible, left in the Tower gate house, thought to have been written the night before he died, were words of faith and trust. It was a verse of his own bitter-sweet poem, remembered from the courting of his wife, now with slight amendment and the addition of a new last couplet:

> Even such is Time, which takes our trust
> Our Youth, our joys and all we have,
> And pays us but with age and dust;
> Who in the dark and silent grave,
> When we have wandered all our ways,
> Shuts up the story of our days,
> But from which earth and grave and dust
> The Lord shall raise me up, I trust.

It was wholly in tune with the Passionate Pilgrim poem, written in expectation of death at the time of his first trial at Winchester. It reflects his words with his spiritual counsellor, the Dean of Westminster. It echoes the majestic concept of the Divine or providential nature of history in his *History of the World*. His scaffold speech, with its humility, penitence and stark recognition of the vanities and follies of his past life, was a dignified and noble acceptance of his fate.

The last act was also, I believe, an only partly-conscious 'psychodrama'—a re-enacted trauma—of striving towards wholeness. In his last days he fulfilled the motto he first served under: 'Let valour end my life.' And that valour included the physical, mental and spiritual elements of his being. The fears, furies and follies were all conquered, quieted and assimilated. In spite of everything he trusted that he would be accepted at his latter end, into life eternal, by the blood of Christ.

One biographer has said that everything he touched seemed to glow, but when he withdrew it seemed to fade. If he was not a torchbearer he was certainly a lamp-lighter and many lamps have been lit from the example of his glowing vision. He had given full expression to all his parts: courtier, colonizer, soldier, ship-designer, explorer, verse and prose writer, parliament man, patron of the arts and sciences, falconer, landscape gardener, botanist, chemist, historian, war reporter by land and sea, war strategist, bibliophile and antiquarian. But I believe his greatest achievement was his finding of psychological wholeness and the transcendence he so passionately desired.

NOTES

1. David B. Quinn (ed.), *The Voyages and Colonizing Enterprises of Sir Humphrey Gilbert*, Hakluyt Society, 1940, pp. 40–5.
2. W. G. Gosling, *Life of Sir Humphrey Gilbert*, Connecticut 1970, p. 150.
3. *Calendar of State Papers Spanish 1568–79*, no. 598.
4. C. E. Champernowne, *The Champernowne Family*, 1954 and Champernowne Family Records.
5. E. L. Smith (ed.), *The Penguin Book of Elizabethan Verse*, 1965.
6. Agnes M. C. Latham, *The Poems of Sir Walter Raleigh*, 1951, p. xxvii.
7. Edward Edwards (ed.), *Life of Sir Walter Raleigh*, 1868, vol. II, p.384.
8. E. C. Wilson, *Prince Henry and English Literature*, Cornell University Press, 1946, pp. 181–2, 185.
9. *Complete State Trials*, ed. W. Cobbett, vol. II, col. 40.

7

John White and his Drawings of Raleigh's Virginia

Paul Hulton

John White, artist and governor—artist of the first Roanoke colony, 1585–6, governor of the second colony of 1587. About his governorship and its aftermath White left two accounts, published by Richard Hakluyt,[1] but of his career as an artist he has left us no information whatever and almost nothing is known about his life. In contrast to such obscurity are his extraordinarily precise and accurate drawings of the native inhabitants and natural life, as well as his maps, of Raleigh's Virginia (part of the modern states of North Carolina and Virginia), overwhelmingly the most valuable graphic evidence about the New World to survive from the sixteenth century.[2]

When White was appointed governor of the 'Cittie of Raleigh' in 1587, he was described as 'of London, Gentleman'. This is not to be taken in the full literal sense. 'Of London' is likely to have meant only that when he was given that office he had been domiciled in London for some years. Equally the 'Gentleman' seems not to have meant that he came from that class of society. The evidence, such as it is, strongly suggests that he was of modest artisan stock since we know that his daughter had married a bricklayer and tiler before they both sailed with him as colonists in 1587. The title of gentleman was bestowed on him probably because it was considered appropriate for the authority he would exercise as governor.

HIS LIFE—A SKETCH

As a grandfather in 1587—his daughter Eleanor's child, Virginia, was born in the colony in August of that year—White's own birth may be placed in the decade 1540–50. Where he was born, or who his family were, are matters for speculation since, so common is the name, that a number of John Whites compete for identification as the artist. The most likely candidate is a John White of Truro, one of twin brothers.[3] Certainly a Cornish connection would be consistent with the strongly West Country involvement in the earliest English colonizing ventures.

There is equally no certain evidence about his early career as an artist. However, the brief mention of a John White as a member of the Painter-Stainers' Company of London in 1580 could well refer to him.[4] The quality of White's drawings suggests that he was unlikely to have been an amateur, rather that he received a conventional training as an artist apprentice under a master in the appropriate city company—that of the Painter-Stainers, who controlled every sort of painting activity. His later ability as a portraitist in miniature also suggests that his master was a limner or miniaturist.

Other evidence of White's activity as an artist is circumstantial but strong. Among his drawings are portraits of Eskimos as convincing and detailed as any of his Indian portraits. In theory these could be copies, just as his Indians of Florida, a man and a woman, were copied from the work of his older contemporary, the Huguenot artist Jacques Le Moyne de Morgues. But the existence of another Eskimo drawing in a volume of early copies from White suggests otherwise. It is a landscape showing a fight between Eskimos and a boatload of Englishmen.[5] The Eskimos are on a cliff top firing arrows on the English below. The minutely-drawn details of Eskimo arms, dress and equipment, showing how their tents were constructed and their kayaks were paddled among the ice-floes, are too convincing to have been drawn at second hand. Even if it were an excellent copy, the original must surely have been made by one who was an eye-witness. So it would seem that White himself was a member of one of Frobisher's expeditions to the North West. An incident on his second voyage of 1577, off South Baffin Island, is the likely subject of the drawing. On that same voyage an Eskimo man and a woman with her baby were captured and brought back to Bristol. These also were the subjects of White's portraits. There is therefore a case for supposing that White was a experienced explorer-artist in North America before he sailed to Raleigh's Virginia.

There is also good reason to believe that the voyage of 1585 was not White's first to these coasts. As long ago as 1925 the art-historian, Laurence Binyon, pointed out that since in 1593 White wrote to Richard Hakluyt from Ireland that the 1590 voyage to Virginia was his fifth and last, then that of 1585 could not have been his first, his other voyages being fully documented.[6] This could only mean, if White's statement is correct, that he was with Amadas and Barlowe on the reconnaissance voyage to Roanoke Island of 1584. Thus when White was recruited as official artist of the 1585 expedition he was already knowledgeable about conditions in Raleigh's Virginia and experienced in North American exploration.

It is a curious fact that John White's name does not appear on the list of colonists published by Richard Hakluyt for the 1585 expedition. He is mentioned for the first time in the *Tiger*'s journal for 11 July when he is cited as one of a party of senior men setting out from Wococon Island across Pamlico Sound to explore the mainland.[7] It is also the last mention of him, but we can be sure that he was fully occupied drawing and surveying in different parts of the colony until the colonists left Roanoke Island in July 1586.

During White's year in the colony and after his return to England his commitment to the cause of colonization, and his responsible activity, must have made a considerable impression on Raleigh and the promoters of the venture—to such an extent that Raleigh devolved on him and twelve assistants the organization and direction of the second colony. On 7 January 1587 White was named governor of the 'Cittie of Raleigh in Virginia' and with his assistants formed a corporate body. This company was directly concerned with attracting investment in the colony and with recruiting settlers. White's small squadron consisting of the *Lion*, a ship of 120 tons, and two unnamed smaller vessels—a sea-going pinnace and a fly-boat—sailed from Portsmouth on 26 April. Eighty-four men, seventeen women and twelve children made up the number of would-be settlers.

Among them was White's daughter, Eleanor, and her husband, Ananias Dare.

Differences of opinion soon developed between White as captain and Simon Fernandes, the Portuguese navigator of the *Lion*. They came to a head when the crew refused to sail the ship from Roanoke Island north to Chesapeake Bay to search out a suitable site for the colony near a deep-water harbour, in accord with Raleigh's instructions to White. Already it had become clear that White lacked the authority fully to carry out Raleigh's intentions, either as captain of the *Lion* or as governor of the colony. In little more than a month he was persuaded by the colonists to return to England to negotiate further supplies.[8]

The difficult voyage home, when the *Lion* went in search of prizes off the Azores, was symptomatic of future difficulties. Though White had the sympathetic ear of Raleigh, who intended to send out further supplies as speedily as possible, the threat of the Armada thwarted the best efforts of the two men. Plans for Grenville to sail with reinforcements and supplies were cancelled by the Privy Council on 31 March 1588. Though Drake allowed two small vessels, not needed for the defence of the realm, to undertake the voyage the following month, the larger, the *Brave* of 30 tons, turned aside to search for prizes, then was herself crippled and looted by two French ships. There were casualties and White was wounded and forced to return.

Nothing could be done to relieve the colony that year or early in 1589. White persevered in obtaining more merchants to invest in the company and there was certainly a continuing interest in the venture, but for reasons not clear to us now supply ships were not able to sail before March 1590. White has left a vivid account of the ill-fated search for the colonists in appalling weather conditions along the Carolina Outer Banks.[9] Seven men were drowned in an attempt to get ashore but the English ships continued north to Roanoke Island. Off its northern shore they saw a fire on land and were encouraged to 'sound with a Trumpet a Call, & afterwards many familiar tunes of songs, and called them friendly', but 'we had no answer.' Going ashore they found the fire was not man made. They also discovered the word CROATOAN carved in 'fayre Capitall letters' on the site of their former settlement, indicating that the colonists had moved to the island of Croatoan further south. White found three of his old chests dug up and ransacked in the fort, 'my bookes torne from the couers, the frames of some of my pictures and Mappes rotten and spoyled with rayne...'. They decided to sail south to Croatoan to continue the search but the weather grew even worse and finally forced them to leave this dangerous coast and return home.

Thereafter, though there were other attempts over the years to find the colonists, White himself was compelled to abandon plans to supply them, though he continued to believe they were still alive. The last certain evidence of him is in the letter he wrote to Hakluyt from 'my house at Newtown in Kylemore', Co. Cork.[10] In Ireland he had resigned himself to his failure: 'I leaue off prosecuting that whereunto I would to God my wealth were answerable to my will...'. He ends by committing the planters to 'the merciful help of the Almighty'.

HIS ACTIVITIES AS EXPLORER-ARTIST, 1585–6

It is likely that White received a set of instructions as to his duties on his recruitment as expedition artist. These have not survived but others have, such as those given to an otherwise unknown artist, Thomas Bavin, for one of the voyages of Sir Humphrey Gilbert in 1582.[11] They envisaged two men working closely together, a painter-surveyor who was to 'drawe to liefe all strange birdes, beastes, fishes, plantes, hearbes, Trees and fruictes...Also...the figures & shapes of men and woemen in their apparell, as also of their manner of wepons in every place as you shall finde them differing'; and an observer to collect and describe the material drawn, so that a full record could be made of the topography, natural resources and Indian life of the land they were prospecting. These instructions could have been tailored precisely for White the 'painter' and for Thomas Harriot the 'observer', White's quick eye and hand recording every detail, Harriot, with his strongly scientific bias, identifying, naming, describing and arranging the material portrayed.

In May 1585 when Grenville's flagship, the *Tiger*, entered West Indian waters, White and Harriot began their record. A Spanish source tells us that Englishmen on Puerto Rico were collecting banana plants and making drawings of fruits and trees, no doubt referring to the activities of White and Harriot. Drawings survive of a banana plant and its fruits (plate 3), a mammee apple and a pineapple. They evidently also observed and drew a wide variety of fauna—fishes, birds, reptiles, insects and crustaceans, though no drawings of mammals have survived. There are particularly fine drawings of a land crab and a flamingo.

It was not primarily White's intention to illustrate the activities of the colonists but he does so in two drawings, both of them recording episodes on the island of Puerto Rico. The first shows Grenville, the 'admiral', on horseback, returning with his men from a foraging expedition to a fortified encampment made by the English at Guayanilla Bay. Fresh water is being collected and timber brought in on a carriage to build a pinnace, seen nearly finished. A heron, a duck and land crabs are finely drawn in miniature in shallow water at one side of the camp. An English ship, probably the *Tiger*, is lying offshore.[12] This is a vivid and detailed commentary of events between 11 and 23 May. The other drawing shows an enclosure built with bastions like a fort, on a shore near Cape Rojo, surrounding two salt mounds which the English are busy removing from Spanish possession into an English boat.[13] Of special interest is the similarity of the 'fort' to the one now excavated on Roanoke Island which the English were to build there to protect their base.

Towards the end of June the colonists made their first landfall on North American soil, on the island of Wococon in the Outer Banks. The work of exploration, mapping and recording Raleigh's Virginia then began in earnest. Within a very short time of 11 July, when White was mentioned as one of the party about to cross to the mainland, the bulk of the Indian drawings had probably been made, for they quickly discovered the villages of Pomeiooc and Secoton which provided most of White's known Indian portraits and scenes of Indian life. They were for the time being on the best of terms with the Indians, in whom Harriot was particularly interested. His first objective, it seems, was to discover all he could about the tribes which we now know as the Carolina Algonquians. He had already learnt something of their language from the two Indians brought back by the reconnaissance expedition of the previous year. Now was the moment to study the Indians in depth. White's Indian drawings, with Harriot's notes as published in the illustrated edition of his *Briefe and true report of the new found land of Virginia* (1590), show how successful their combined efforts were.[14]

While the Indians were the first priority, the recording of natural life evidently proceeded with as little interruption as possible. Under the heading 'Of Foule' in the *Briefe and true report* Harriot wrote,

> we haue taken, eaten, & haue the pictures as they were there drawne with the names of the inhabitants [i.e. the Indian names], of severall strange sorts of water foule, eight and seventeene kinds more of land foul, although we haue seene and eaten many more, which for want of leasure there for the purpose coulde not be pictured...

It is clear that there was insufficient time for White to draw every species they found. It also seems, to judge by other remarks, that Harriot obtained from the Indians information about fauna and flora which they did not catch or find. As we shall see later, only part of the record survived, although it was evidently their aim to draw every creature and plant they found, and to give each its Indian name (with, if possible, an English equivalent) and description, evaluating its worth for food, medicine, clothing, housing or trade. White's drawings and Harriot's descriptions are the first of such records of the New World to survive of which the word scientific can justly be used. The drawings were derived from a great number of field-sketches made in pen and ink or black lead. Though these have disappeared they must have been sufficiently supplied with colour notes, or colours actually used, to enable White to produce the finished watercolours later in England. It is even possible that some of these were made in the colony, more likely on the return voyage, but the surviving originals were almost certainly made later in England.

White and Harriot's other main task was to survey and map the new land. Both were with the party that explored Pamlico Sound and were probably with the Indians further north in August, in the tribal area of Weapemeoc. It is equally likely that they were in the party sent by Ralph Lane during the winter to explore the south-eastern shore of Chesapeake Bay. In any event the large stretch of land from south of Pamlico Sound to Chesapeake Bay was mapped with extraordinary accuracy, unprecedented in the sixteenth century.[15] There is little doubt that Harriot's expertise accounted for the accuracy of White's map of Raleigh's Virginia, one of only two maps that survive in the original. But neither has the kind of information a trained cartographer would have added—magnetic data, co-

ordinates, soundings and ecological symbols. They are largely outlines garnished with English shipping, Indian canoes, whales, myth-
ical sea creatures and the occasional recognizable fish specimen. The
areas are carefully inscribed with Indian names and the location of
Indian villages marked. Rather more of the experienced map-maker's
detail, it is true, is found in Theodore de Bry's engraving of 'Virginia'
which has upland areas indicated inland to the west and Indian
villages marked with symbols of circular palisading. The engraving
of 'The arriual of the Englishemen in Virginia' goes even further,
showing that White added symbols for crops, wrecks, shoals and
fish-weirs. This map, of which the original is lost, covers the small
area around, and to the north of, Roanoke Island which the colonists
knew best.

The accuracy of these 'Virginia' maps can only have been achieved
by constructing the final map from field sheets, drawn from a survey
employing astronomical observations and a uniform scale, and fitted
together by means of accurate registration and adjustment. White
and Harriot could not possibly have surveyed the whole area. They
would have had to rely on second-hand information for the parts they
did not reach—south of the Pamlico River and areas north of the
south-eastern coastline of Chesapeake Bay. Here and in other periph-
eral parts there are inaccuracies and distortions.

White's other manuscript map covers a much larger area—from
Cape Florida to Chesapeake Bay.[16] Here he had the problem of
reconciling at least three maps of different scales and origins: his own
map of 'Virginia'; a map of Florida evidently supplied by Jacques Le
Moyne de Morgues—with whom he exchanged many ideas—and an
older Spanish map with which he attempted to link the other two.
This accounts for the gross distortion of the central area and other
inconsistencies. We can be sure that Harriot was not concerned in the
compilation of this map which must have been put together after the
return of the colonists to England in late July 1586.

After a year's extremely active surveying and recording in the
colony White and Harriot had armed themselves with a wide range
of information about the Carolina Algonquians, as yet unaffected by
European influences, new species of flora and fauna, some of which
held the promise of wealth and trade, and sketch-maps covering a
considerable part at least of the coastal areas of the new land, its river

estuaries and offshore islands. But when Drake was in the last stages of repatriating the colonists, on 18 July 1586, near disaster struck. A hurricane had blown up and as the pinnaces sent to Roanoke Island to take off the baggage party were leaving they ran aground in high seas so that 'most of all wee had,...all our Cardes [maps], Bookes and writings, were by the Saylers cast over boord...'[17] How much of White's record was lost we shall never know but even allowing for exaggeration by Lane it must have been quite a large part. What has survived may have been based in the main on drawings shipped home at an earlier stage.

THE DRAWINGS

From *A briefe and true report* it is clear that Harriot and White intended to publish an illustrated account of the colony from their own 'chronicle'. Either the plan was too ambitious or the loss of material when they left the colony could not be made good. What Harriot published in 1590, the De Bry edition of his *Briefe and true report*, was an abstract of the intended account, with engravings of much of his Indian material only, in addition to the map of Raleigh's Virginia and 'The arriual of the Englishemen' mentioned above. This famous book was the first part of De Bry's *America* and the second edition of Harriot's *Report*, first issued in 1588. But the drawings De Bry collected for publication were not the same as those of the set that has survived. The ones De Bry used are more elaborate and detailed, often with back views of the main figures added to show off their dress, body ornament or hairstyle, and frequently with landscape backgrounds. The surviving drawings are in most instances the basic composition, without frills. For this reason they could have been White's own set kept for reference, perhaps the one he used when making others for presentation. And it is certain that a number of others were made. Harriot most obviously would have required a set for his own use. Then there are likely to have been presentation sets for the leading figures in the colonizing circle, Raleigh almost certainly, Sir Thomas Walsingham as the most powerful investor in the 1585 enterprise, probably Grenville as admiral of the 1585 expedition. Then there was the Queen herself, somewhat cautiously involved in the venture, who would certainly have been the recipient of a set if Raleigh had considered it appropriate. White also gave replicas of individual drawings to people who, for some particular reason, wanted them—Thomas Penny, the entomologist, for example, who received from White drawings of West Indian fireflies and a gadfly, and John Gerard, author of the *Herball*, to whom White gave at least one drawing, the milkweed.[18]

The seventy-five drawings in watercolour are curiously mixed in subject matter, for apart from the Eskimo portraits and the Indians of Florida already mentioned, there are five figures of Picts and Ancient Britons and five assorted orientals, as well as two European birds, none of which has the least to do with the Roanoke colony. This perhaps is another reason why this set of drawings may be considered as part of White's collection rather than a presentation set. In making this collection of races White was in line with other contemporary artists who reflected a growing interest in the characteristics of different cultures both of the Old World and the New.

Mention has been made of a volume of early copies of White's drawings in which the landscape with Eskimos fighting Englishmen is found. These copies, though of much lower artistic quality and scientific accuracy (apart from this one drawing) than the original set are important in extending our knowledge of White's work, including a considerable number of birds and fishes of Raleigh's Virginia absent in the originals.[19] Here too are costume studies, different ones, entirely unrelated to the Roanoke venture.

The volume of copies was discovered by Dr Hans Sloane in about 1706, in the hands of White's descendants.[20] He at once recognized that the Indian subjects were closely related to the De Bry engravings. Not unnaturally he believed them to be their originals for White's original set was not to re-emerge to public knowledge for many years. They are in fact very early seventeenth-century copies or may even have been made in the last years of the sixteenth century. They are presumably the work of members of White's family (several hands seem to have been involved), perhaps of the next generation to John White. Though twenty-seven of them more or less duplicate subjects in the original set, there are forty-four drawings of birds, fishes and reptiles of 'Virginia' taken from originals now lost. Sloane was eventually able to acquire them and they are now known as the Sloane copies. He placed so much importance on them that he had many of them duplicated and reduplicated for the use of scholars and students. It is interesting to note that the Sloane copies reveal differences in detail from both the original drawings and the De Bry engravings, showing that they were taken from a different set of drawings now lost, perhaps the basic collection of assorted drawings still available to the copyists within the family.

The techniques which White used as an artist are of special interest. Whereas most of his contemporaries in northern Europe used strongly drawn outlines, usually in pen, to enclose their figures, which they then coloured in with the brush, White seems to have done most of his drawing with the point of the brush. There are certainly indications of faint black lead outlines but he used these as guide-lines only. Though his use of dense, opaque and metallic pigments was traditional, his application of watercolour washes to unprepared paper, often using it to obtain highlights, was an original feature which seems almost to anticipate much later developments in the English school of watercolour. Though he was normally careful, precise and concerned with naturalistic accuracy (yet he conceived his images still somewhat in the Mannerist tradition of his contemporaries in northern Europe), especially for example in his successful attempts to simulate the texture of deerskin, of turtleshells and of rush matting, he could use his brush spontaneously and effectively, and with sensitivity, to draw small figures of animals, birds and fishes. The accuracy of his work was no doubt to some extent disciplined by Harriot, the scientist, but apart from his workmanlike competence he often shows touches of originality which indicate not only his commitment to his duties as expedition artist but also suggest the pleasures of discovery.

The original drawings cannot be precisely dated but were clearly made within a short time of each other to judge from the uniformity of style and paper. Being the simple, basic drawings, they were probably already finished before Theodor de Bry collected his set of drawings on his visit to England in 1588. Perhaps it is not surprising that their author, who never adds his signature, was forgotten, whereas De Bry, his publisher and engraver, was not, for his Indian plates remained alive in the European consciousness long after the Carolina Algonquians had become extinct. Indeed for much longer than a century they were taken as the type of the North American Indian (in particular the Indian chief painted and armed—plate 4) when in fact they portrayed only one sub-section of a group of Indians quite distinct from and not nearly as numerous as many others across North America.

The influence of White's drawings was long delayed and sometimes indirect. Their very considerable importance for ethnography and natural history has only been fully realized in comparatively recent years and is still being worked out. Yet there was some historical influence. Linnaeus, the great innovator in the classification of species, refers to a number of illustrations which White originated without realising who their author was. In his *Systema naturae* (10th edn 1758) he cites Mark Catesby's *Natural history of Carolina, Florida and the Bahama Islands* (1731–43) in five instances and for two of his plates. These five subjects were among seven copied by Catesby without acknowledgment, except in one instance, directly from the Sloane copies.[21] In this way one of the greatest figures of the eighteenth-century Enlightenment paid indirect though powerful tribute to John White, an artist of whom he had never heard.

NOTES

1. Transcribed and carefully analysed in David Beers Quinn (ed.), *The Roanoke Voyages 1584–90* (Hakluyt Society, 1955), pp. 515–38, 598–622.
2. Reproduced and described in Paul Hulton, *America 1585: the complete drawings of John White* (University of North Carolina Press and British Museum Publications, 1984).
3. J. L. Vivian, *Visitations of Cornwall* (1887), p. 553.
4. See W. A. D. Englefield, *The history of the Painter Stainers Company of London* (1923), p. 53n.
5. See Hulton, *America 1585*, fig. 43 and p. 194.
6. Laurence Binyon, 'The drawings of John White', *Walpole Society*, XIII, 1924–5 (Oxford 1925), p. 20.
7. Quinn, *Roanoke Voyages*, p. 190.
8. *Ibid.*, pp. 534–5.
9. *Ibid.*, pp. 612–18.
10. *Ibid.*, pp. 712–16.
11. See Hulton and Quinn, *The American drawings of John White* (London and Chapel Hill 1964), I, pp. 34–5.
12. Hulton, *America 1985*, pl. 3 and p. 173.
13. *Ibid.*, pl. 4 and pp. 173–4.
14. For an accessible modern facsimile see that issued by Dover Publications Inc. with an introduction by P. Hulton, New York, 1972.

15. See Quinn, *Roanoke Voyages*, p. 461, map 7, pp. 842–72, map 12; Hulton, *America 1585*, pp. 32–3, 183, pl. 60.
16. See Quinn, *Roanoke Voyages*, pp. 460, 848, map 6; Hulton, *America 1585*, pp. 33–4, 183, pl. 59.
17. Quinn, *Roanoke Voyages*, p. 293.
18. Hulton, *America 1585*, pp. 20–21.
19. Reproduced *ibid.*, figs 34–106, described pp. 186–206.
20. Quinn, *Roanoke Voyages*, pp. 394–5; Hulton, *America 1585*, p. 21.
21. Hulton, *ibid.*, p. 32.